T0134409

Intelligent Systems Reference Library

Volume 180

The aim of this series is to publish a Reference Library, including novel advances and developments in all aspects of Intelligent Systems in an easily accessible and well structured form. The series includes reference works, handbooks, compendia, textbooks, well-structured monographs, dictionaries, and encyclopedias. It contains well integrated knowledge and current information in the field of Intelligent Systems. The series covers the theory, applications, and design methods of Intelligent Systems. Virtually all disciplines such as engineering, computer science, avionics, business, e-commerce, environment, healthcare, physics and life science are included. The list of topics spans all the areas of modern intelligent systems such as: Ambient intelligence, Computational intelligence, Social intelligence, Computational neuroscience, Artificial life, Virtual society, Cognitive systems, DNA and immunity-based systems, e-Learning and teaching, Human-centred computing and Machine ethics, Intelligent control, Intelligent data analysis, Knowledge-based paradigms, Knowledge management, Intelligent agents, Intelligent decision making, Intelligent network security, Interactive entertainment, Learning paradigms, Recommender systems, Robotics and Mechatronics including human-machine teaming, Self-organizing and adaptive systems, Soft computing including Neural systems, Fuzzy systems, Evolutionary computing and the Fusion of these paradigms, Perception and Vision, Web intelligence and Multimedia.

** Indexing: The books of this series are submitted to ISI Web of Science, SCOPUS, DBLP and Springerlink.

More information about this series at http://www.springer.com/series/8578

Valentina E. Balas · Vijender Kumar Solanki ·
Raghvendra Kumar
Editors

Internet of Things and Big Data Applications

Recent Advances and Challenges

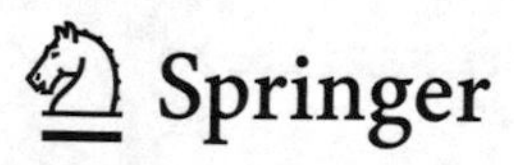

Editors
Valentina E. Balas
Department of Automatics
and Applied Software
Aurel Vlaicu University of Arad
Arad, Romania

Vijender Kumar Solanki
Department of Computer Science
and Engineering
CMR Institute of Technology (Autonomous)
Hyderabad, India

Raghvendra Kumar
Department of Computer Science
and Engineering
GIET University
Gunupur, India

ISSN 1868-4394 ISSN 1868-4408 (electronic)
Intelligent Systems Reference Library
ISBN 978-3-030-39121-8 ISBN 978-3-030-39119-5 (eBook)
https://doi.org/10.1007/978-3-030-39119-5

This Springer imprint is published by the registered company Springer Nature Switzerland AG
The registered company address is: Gewerbestrasse 11, 6330 Cham, Switzerland

Preface

This comprehensive and timely publication aims to be an essential reference source, building on the available literature in the diverse field of IoT and big data while developing applications using cloud computing, soft computing, programming paradigms, etc. The objectives of this book are to identify different issues, suggest feasible solutions to those identified issues, and enable researchers and practitioners from both academia and industry to interact with each other regarding emerging technologies related to IoT and big data. In this book, we look for novel chapters that recommend new methodologies, recent advancement, system architectures, and other solutions to prevail over the limitations of IoT and big data.

Chapter 1 proposes a dynamically distributed algorithm for processing the job scheduler process that allocates data processing-based workload formation for individual nodes to increase their performance related to the throughput. The relative simulations are implemented using 2.4 GB weather temperature conversion datasets demonstrates that our proposals can significantly reduce processing costs though prioritizing working nodes enhanced compared to previous methodologies.

Chapter 2 discusses about classification model is trained which directs source language to target language on the basis of translation knowledge and parameters defined. The algorithm adopted here is expectation–maximization algorithm which removes ambiguity in parallel corpora by aligning source sentence to target sentence. It considers possible translations from source to target language and selects the one that fits the model on the basis of bilingual evaluation understudy (BLEU) score. The only requirement of this learning is unlabeled data in the target language. The algorithm can be evaluated accurately by running a separate classifier on different parallel corpora. We use monolingual corpora and machine translation in our study to see the effect of both the models on our parallel corpora.

Chapter 3 addresses the problems which are faced in the healthcare sector. The interview methodologies with every kind of questions are examined and find the solution for this problem.

Chapter 4 designs a self-contained evaluation of ML techniques and IoT applications in intelligent transportation systems (ITS) and attain a clear view of the developments in the above-mentioned fields and spot possible coverage musts.

From the reviewed articles, it becomes insightful that there is a possible lack of ML coverage for the smart lighting systems and smart parking applications. Additionally, route optimization, parking, and accident/detection tend to be the most popular ITS applications among researchers, henceforth there is a huge possibility in implementing the IoT with real-world applications with the support of various big data concepts and machine learning algorithms along with blockchain technology.

Chapter 5 presents the classic wireless IoT application scenarios, along with associated models of safety and privacy assault. This paper further presents recent advances in wireless IoT privacy conservation mechanisms along with classification of the application scenarios. In this paper, open problems and future research directions are addressed for the implementation situations in wireless IoT.

Chapter 6 discusses Internet of Things (IoT) in health care which is a revolution in patient's care with improved diagnosis, real monitoring and preventive as well as real treatments. The IoT in health care consists of sensors-enabled smart devices that accurately collect data for further analysis and actions. By using real-time data, the devices allowed tracking, monitoring, and management in order to improve health care.

Chapter 7 focuses on the review of the mobile applications available in the Google Play Store that are dedicated to cardiac patients. The number of cardiac patients is increasing, but there are no mobile applications that aid cardiac patients by providing monitoring of different parameters, including the calorie intake and the calories burned. However, the mobile applications that can be adapted to this type of people were analyzed. We found six notable mobile applications. Their features can be grouped in diet, anthropometric parameters, and physical activity.

Chapter 8 discusses more about ML and DM techniques, and we can detect those intrusions and stop them from occurring. This review contains information about how intrusion detection methods in cyber security can be used with machine learning and data mining techniques.

Chapter 9 proposes a system based on machine learning for classification of breast cancer (BC) along with the comparative study of two machine learning (ML) classifiers. The idea is to select the region of interest (ROI) at very first from the mammograms. At that point, important features have been extracted using gray-level co-occurrence matrix (GLCM). Thereafter, extracted features are then utilized to train our classifiers SVM and KNN individually. The mammogram is then characterized either into benign or malignant using the trained classifier. Proposed system is implemented on standard MIAS databases. Classification performance of both classifiers are contrasted in terms of accuracy, recall, precision, specificity, and F1 score. We found that SVM achieved higher accuracy of 94% than KNN with better recall and F1 score.

In Chap. 10, survey of various MAC protocols for energy harvested wireless sensor network is presented and it is based on different parameters such as energy efficiency, end-to-end delay, Quality of service, and scalability. Survey is presented in the order of proposed year so one can easily find various gaps in the subsequent studies and get the idea of future research.

In Chap. 11, exhaustive parametric investigation has been done to comprehend the impacts of different dimensional parameters and to streamline the execution of the antenna. A substrate of low dielectric consistent is chosen to get a minimal transmitting structure that meets the requesting data transfer capacity determination. The reflection coefficient at the contribution of the upgraded rectangular formed microstrip patch antenna is underneath—10 dB over the whole recurrence band. The estimation results are in fantastic concurrence with the HFSS 15.v re-enactment results.

Chapter 12 focuses on a magnetic resonance imaging (MRI) reconstruction that gives brisk realization. This diminishes the scanning cost and image reconstructed in very fewer time. In this method, generative adversarial network (GAN) designed a generator which gives the better enhancement like texture smoothness and high resolution.

Chapter 13 provides an overview of the disciplines of artificial intelligence with respect to soft computing and presents the benefits of soft computing over traditional hard computer technologies and their disadvantages.

In Chap. 14, study, different PAPR reduction techniques for filter band multi-carrier with offset quadrature amplitude symbol mapping called as FBMC-OQAM are presented, such as nonlinear companding transform, improved hybrid techniques based on conventional partial transmit sequence (PTS), iterative clipping and filtering (ICF), and tone reservation method.

Chapter 15 distinguishes the IoT-based security mechanism for providing the security which also diminishes the healthcare facilities. The collaboration model minimizes the risk for implementing the e-healthcare policies all over the world to decide the economical and social solutions for the healthcare information systems.

Chapter 16 discusses the criteria used to choose the right predictive model algorithm.

Chapter 17 analyzes specific body parameters which are heart rate (HR), blood pressure(BP), temperature. These parameters are examined with the use of body sensors which are attached to the body. These sensors are interfaced with an uC with other components to help in the proper examination of the parameters of our body. This product is capable of communicating to the end user (qualified person) with the help of ZigBee Wi-Fi, GSM shield by creating a wireless sensor network (WSN) and transmitting data from the sensor to the uC using body area network (BAN).

In Chap. 18, the proposed antenna has perfect broadside radiation pattern and high gain of 6.9 dbi below—10 db. The antenna is simulated using HFSS software at operating frequency of 2.4 GHz and exhibits good result.

In Chap. 19, the proposed system is based on a passive user inertial method, which collects and uploads fingerprints automatically in daily life using MAS. We show how to handle multiple floors and stairways, how to handle symmetry in the environment, and how to initialize the localization algorithm using Wi-Fi signal strength to reduce initial complexity.

Chapter 20 reviews the different biosensors based on IONPs and their applications, IONPs in imaging and drug delivery. Iron oxide nanoparticles can be synthesized with co-precipitation, thermal decomposition, and microwave techniques to achieve optimum results and then coated with polymers and further loaded with related drug, biomolecule, or a dye to obtain an optical result. These IONPs are then doped with other nanoparticles to form composites with nanostructures as graphene or biomolecules as Chitosan are further usually dispersed over a glass or a silicon substrate to fabricate a biosensor. IONPs tend to achieve superparamagnetic property and therefore known as SPIONs that increases the application of these nanoparticles in all spectrums of biomedical application.

Chapter 21 uses cold-FET and hot-FET techniques to obtain intrinsic and extrinsic parameters to demonstrate the impact of parasite element passivation, parasitic capacitances, resistances, and inductances. We learn the magnitude of their effect on the efficiency of power and microwave from this stage. The validity of the suggested algorithm was carefully checked with an outstanding correlation between the parameters measured and modeled very well.

In Chap. 22, review on digital watermarking techniques is evaluated and described. The watermarking is produced since the image is contented and could be preserved as an arithmetical impression of fingerprint of the image. The changes based on procedure is cast-off to encrypt the evidence in the histogram area that projected watermarking is vigorous sufficient in contradiction of any deprivation, equitation and occurrence.

There have been several influences from our family and friends who have sacrificed lot of their time and attention to ensure that we are kept motivated to complete this crucial project. The editors are thankful to all the members of Springer (India) Private Limited especially Prof. (Dr.) Lakhmi C. Jain and Aninda Bose for the given opportunities to edit this book.

Arad, Romania — Valentina E. Balas
Hyderabad, India — Vijender Kumar Solanki
Gunupur, India — Raghvendra Kumar

About This Book

With the progress of sensors and wireless technology, the Internet of Things (IoT)-related applications are gaining much attention. With more and more devices are getting connected, they become the potential components of some smart applications. Thus, a global enthusiasm has sparked over various domains such as health, agriculture, energy, security, and retail. So, in this book, the main objective is to capture this multifaceted nature of IoT and big data in one single place. Each of the chapters of this book will cover a domain that is significantly impacted by the growth of soft computing. According to the contribution of each chapter, the book will also provide a future direction for IoT and big data research.

Key Features

1. Covering the potential impactful growth of IoT system and big data.
2. Contributors belong to different parts of the world working with multidisciplinary laboratory will be supporting us by contributing their research work in our book
3. Contains insightful approach for interdisciplinary approach of IoT, e.g., energy saving devices, sensors-enabled environment, IoT, etc.,
4. Presents several chapters' emphasis on improving the efficiency and growing deed through IoT/big data approach
5. Exploration of cutting-edge technologies through sensor-enabled environment for industry.

Contents

About the Editors

Valentina E. Balas, Ph.D. is currently Full Professor in the Department of Automatics and Applied Software at the Faculty of Engineering, "Aurel Vlaicu" University of Arad, Romania. She holds a Ph.D. in Applied Electronics and Telecommunications from Polytechnic University of Timisoara. Dr. Balas is the author of more than 270 research papers in refereed journals and international conferences. Her research interests are in intelligent systems, fuzzy control, soft computing, smart sensors, information fusion, modeling and simulation. She is Editor in Chief to *International Journal of Advanced Intelligence Paradigms* (IJAIP) and to *International Journal of Computational Systems Engineering* (IJCSysE), member in Editorial Board member of several national and international journals, and is evaluator expert for national and international projects. She served as General Chair of the International Workshop Soft Computing and Applications in seven editions 2005–2016 held in Romania and Hungary. Dr. Balas participated in many international conferences as Organizer, Session Chair, and member in International Program Committee. Now, she is working in a national project with EU funding support: BioCell-NanoART = Novel Bio-inspired Cellular Nano-Architectures—For Digital Integrated Circuits, 2M Euro from National Authority for Scientific Research and Innovation. She is a member of EUSFLAT, ACM and a Senior Member IEEE, member in TC—Fuzzy Systems (IEEE CIS), member in TC—Emergent Technologies (IEEE CIS), and member in TC—Soft Computing (IEEE SMCS). Dr. Balas was Vice-president (Awards) of IFSA International Fuzzy Systems Association Council (2013–2015) and is Joint Secretary of the Governing Council of Forum for Interdisciplinary Mathematics (FIM), A Multidisciplinary Academic Body, India.

Vijender Kumar Solanki, Ph.D. is an Associate Professor in Computer Science & Engineering, CMR Institute of Technology (Autonomous), Hyderabad, TS, India. He has more than 10 years of academic experience in network security, IoT, Big Data, Smart City and IT. Prior to his current role, he was associated with Apeejay Institute of Technology, Greater Noida, UP, KSRCE (Autonomous) Institution, Tamilnadu, India and Institute of Technology & Science, Ghaziabad, UP, India.

He is member of ACM and IEEE. He has attended an orientation program at UGC-Academic Staff College, University of Kerala, Thiruvananthapuram, Kerala & Refresher course at Indian Institute of Information Technology, Allahabad, UP, India. He has authored or co-authored more than 50 research articles that are published in various journals, books and conference proceedings. He has edited or co-edited 14 books and Conference Proceedings in the area of soft computing. He received Ph.D in Computer Science and Engineering from Anna University, Chennai, India in 2017 and ME, MCA from Maharishi Dayanand University, Rohtak, Haryana, India in 2007 and 2004, respectively and a bachelor's degree in Science from JLN Government College, Faridabad Haryana, India in 2001. He is the Book Series Editor of Internet of Everything (IoE): Security and Privacy Paradigm, CRC Press, Taylor & Francis Group, USA; Artificial Intelligence (AI): Elementary to Advanced Practices Series, CRC Press, Taylor & Francis Group, USA; IT, Management & Operations Research Practices, CRC Press, Taylor & Francis Group, USA; Bio-Medical Engineering: Techniques and Applications with Apple Academic Press, USA and Computational Intelligence and Management Science Paradigm, (Focus Series) CRC Press, Taylor & Francis Group, USA. He is Editor-in-Chief in *International Journal of Machine Learning and Networked Collaborative Engineering* (IJMLNCE) ISSN 2581-3242; *International Journal of Hyperconnectivity and the Internet of Things* (IJHIoT), ISSN 2473-4365, IGI-Global, USA, Co-Editor Ingenieria Solidaria Journal ISSN (2357-6014), Associate Editor in *International Journal of Information Retrieval Research* (IJIRR), IGI-GLOBAL, USA, ISSN: 2155-6377 | E-ISSN: 2155-6385. He has been guest editor with IGI-Global, USA, InderScience & Many more publishers. He can be contacted at spesinfo@yahoo.com or vijendersolanki@ieee.org.

Dr. Raghvendra Kumar is working as Associate Professor in Computer Science and Engineering Department at GIET University, India. He received B. Tech, M.Tech, and Ph.D. in Computer Science and Engineering, India, and Postdoc Fellow from Institute of Information Technology, Virtual Reality and Multimedia, Vietnam. He serves as Series Editor Internet of Everything (IOE): Security and Privacy Paradigm, Green Engineering and Technology: Concepts and Applications, publishes by CRC press, Taylor & Francis Group, USA, and Bio-Medical Engineering: Techniques and Applications, Publishes by Apple Academic Press, CRC Press, Taylor & Francis Group, USA. He also serves as acquisition editor for Computer Science by Apple Academic Press, CRC Press, Taylor & Francis Group, USA. He has published number of research papers in international journal (SCI/SCIE/ESCI/Scopus) and conferences including IEEE and Springer as well as serve as organizing chair (RICE-2019, 2020), volume Editor (RICE-2018), Keynote speaker, session chair, co-chair, publicity chair, publication chair, advisory board, technical program committee members in many international and national conferences and serve as guest editors in many special issues from reputed journals (Indexed By: Scopus, ESCI, SCI). He also published 13 chapters in edited book published by IGI Global, Springer, and Elsevier. His researches areas are computer networks, data mining, cloud computing and secure multiparty computations, theory of computer science and design of algorithms. He authored and edited 23 computer science books in field of Internet of Things, Data Mining, Biomedical Engineering, Big Data, Robotics, and IGI Global Publication, USA, IOS Press Netherland, Springer, Elsevier, CRC Press, USA. He is Managing Editor in International *Journal of Machine Learning and Networked Collaborative Engineering* (IJMLNCE) ISSN 2581-3242.

Chapter 1
Optimized Cost Reduced Parallel Processing Methodology for Implementing Effective Big Data Analytics

Y. Harold Robinson, E. Golden Julie and S. Balaji

Abstract The concept of MapReduce is utilized for efficient big data processing current days. The dissimilar kind of flavours is utilized with Apache Hadoop for big data processing. The Hive software is combined with the Java advanced features for implementing the Hadoop operation to generate the efficient development of big data processing. The parallel processing can be done effectively using the MapReduce concept of Hadoop technique. Most of the big data processing across the world is implementing this concept in efficient manner. We propose a dynamically distributed algorithm for processing the job scheduler process that allocates data processing based workload formation for individual nodes to increase their performance related to the throughput. The relative simulations are implemented using 2.4 GB weather temperature conversion datasets demonstrates that our proposals can significantly reduce processing costs though prioritizing working nodes enhanced compared to previous methodologies.

Keywords MapReduce · Distributed programming · Apache Hadoop · Big data · Hadoop distributed file system

1.1 Introduction

The term MapReduce was first coined by Google when they first published a study [1] that specifies the implementation required to initiate a paradigm to process large

Y. Harold Robinson (✉)
School of Information Technology and Engineering, Vellore Institute of Technology, Vellore, India
e-mail: yhrobinphd@gmail.com

E. Golden Julie
Department of Computer Science and Engineering, Anna University Regional Campus, Tirunelveli, India
e-mail: goldenjuliephd@gmail.com

S. Balaji
Department of Computer Science and Engineering, Francis Xavier Engineering College, Tirunelveli, India
e-mail: sbalajiphd@gmail.com

V. E. Balas et al. (eds.), *Internet of Things and Big Data Applications*, Intelligent Systems Reference Library 180, https://doi.org/10.1007/978-3-030-39119-5_1

data quickly in a distributed manner. The efficiency of MapReduce is in its ability to deploy and manage programs across domains to build a cohesive unit that can process data and acquire results effectively [2]. Apache Hadoop developed which happens to be a pioneer open source MapReduce implementation. It's flexibility at the core to interact with any of the procedural languages such as C++, Java, Python, Ruby, SQL like via Hive, Piglets to initiate MapReduce has always been its positive feature. Its implementation in prototypes such as Nutch search engine project in 2005 captured everyone's attention but the real deal was when it processed a terabyte of data in 209 s using a 10,000-core Hadoop cluster at Yahoo in April 2008 [3].

Relational Database Management System (RDBMS) aided with Structured Query Language (SQL) are often used to analyze tabular data, they are more suited when the volume of the data is small (<100 MB) and the data in question is continuously updated [4]. The MapReduce approach supplements RDBMS systems especially with regard to structure, format and volume of data and the only difference is the data in question is stored once and accessed many times [5]. The Hadoop based MapReduce approach is very helpful the when the volume of the data is huge and it is unstructured [6]. The growing trend is to use RDBMS systems for dynamic events and activities, for computationally expensive tasks like batch processing of files, using non relational data processing tools such as NoSQL Database or Hadoop etc. is more beneficial. The term Big Data [7] is often associated with Hadoop as companies like Facebook, Yahoo pioneers of MapReduce technology demonstrated its power in processing large volumes of datasets [8].

In this study, we analyze and highlight the performance differences between having Hadoop implementations with or without a processing aware job scheduler [9]. The data set used for processing is a 2.4 GB weather temperature conversion sets and processed in Hadoop Single Node Cluster mode [10]. The performance parameters such as Hadoop chunk size, processing nodes, processing time are considered to validate the performance difference of a processing aware job scheduler and a plain job scheduler [11]. Without a processing aware job scheduler the plain Hadoop implementations fared poorly with respect to the performance parameters. Besides the weather temperature conversion datasets similar datasets from other scientific domains such as environmental analysis [12], energy analysis [13], bioinformatics [14], simulation data [15] can also be used to validate the proposed concept.

MapReduce model is a computational domain that implements distributed processing of big data sets. Most used tasks with respect MapReduce on big data are as follows:

- Distributed globally search a regular expression and print (grep)
- Count frequencies
- Graph structure representation of web artefacts
- Term-frequency
- Inverse document frequency.

The four fundamental stages of any MapReduce model are as follows:

- Input splits
- Split based key value pair generation often called map
- Iterative sort, copy, merge operations on map
- Summing up similar keys with their respective
- Values often called reduce.

The advanced methodology [16] is the first to explore the possibility of performance improvement in the execution of a job by modifying network usage in such a way that it upholds high network resource utilization at reduced network congestion. The latest prototype [17] named Purlieus demonstrated the efficiency of modifying the default MapReduce job assignment system especially using Amazon AWS clouds by using bridge machines between cloud instances and local machines. These partitions reduce overall network congestion using time synchronized shuffling between cloud instances and local machines. But no research was implemented to optimize the shuffling module to reduce network traffic loads in MapReduce tasks. The research with respect to shuffling of data partitions among intermediate workers (data nodes) was picked up [18]. The default configuration of Hadoop is the implementation of a hash map based data structure that would produce in coherent and in-consistent results which require further rounding-off key value pairs to synchronize the input and output data. The overcame the rounding off process with a new scheme called fairness-aware hash-map approach that has a cache to store and audits all the intermediate distributions to guarantee a balanced load distribution among worker nodes [19]. The suggested modifying work allocation scheduler itself and in that regard a scheduler algorithm was developed for data (key value pairs) distributions to enhance load sharing among worker nodes [20]. The improved scheme load balancing with an efficient load balancing scheme of theirs and compared both methods to evaluate skew-ness performance of MapReduce jobs of large datasets [21].

All researches so far have been implemented with respect to task reduction, optimizing shuffling, partitions, generation of key value pairs, local aggregation, cross domain mappers, network aggregation etc. but not on the job scheduler itself which happens to be the current scope of our study.

1.2 Proposed Work

1.2.1 System Model

Flow representation is represented in Fig. 1.1. The problem of optimizing the job scheduler for an efficient distributed processing in MapReduce paradigm was researched earlier. We shall discuss some such methods. Most previous approaches concentrated on utilizing the Hadoop MapReduce platform for their respective domains but some focused on performance improvement within the platform itself especially with respect to tweaking the data transmissions and work allocations between the Hadoop entities.

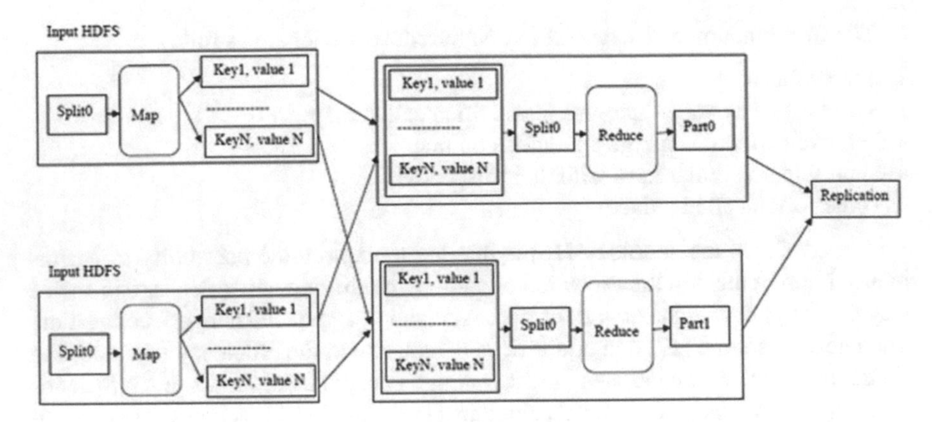

Fig. 1.1 A snapshot of MapReduce process of data flow in a typical Hadoop setup

The following node clusters will be deployed in a usual Hadoop big data processing system. Our proposal of optimizing the job allocation tasks will be initiated at Job Tracker for load aware job assignment among data nodes. Client submits job for execution JobTracker One per Hadoop cluster coordinates job Scheduler TaskTracker One per cluster node executes individual map/reduce tasks MapTask or ReduceTask at Child (Data Nodes) application code architectural representation of the proposed system is as follows in Fig. 1.2.

The default capacity scheduler of Hadoop was intended to permit smooth sharing of chunks of big data to be processed between Hadoop node clusters name node, data node, job tracker, task tracker, data node, job tracker, task tracker, etc.) in an anticipated, quick and straight forward way using pre-defined or default configurations while utilizing the queues provided at each node.

We propose to consider data segregation and conglomeration for a MapReduce work with a motive to minimize total round trips and hence the aggregated system traffic between Hadoop clusters by assigning jobs to more resource centric workers instead of equal distributions. Specifically, we propose an appropriate job queue estimation calculation for big data applications by splitting the primary large data into process able chunks say 128 MB (64 MB Chunk Size for Hadoop) that can be fathomed and processed in parallel at data nodes. Algorithmic representation of the proposed scheduler is as shown in Fig. 1.3.

The proposed scheduler manages the data segments accumulation and assignments in a dynamic way based on the load awareness of processing nodes. Simulated results using Hadoop in pseudo distributed mode shows that our method can altogether decrease job allocation costs. It assures job allocation guarantees for queues and simultaneously provisioning elasticity for queues in the sense that limited capacity of worker nodes determines less possible throughput and a more resource centric worker node determines more throughput. These queues at worker nodes can be

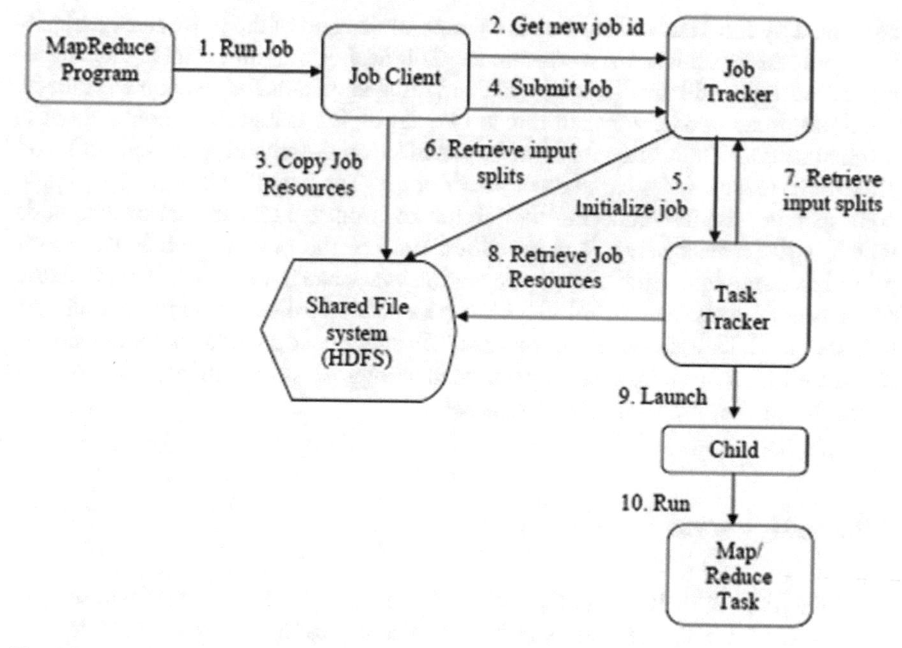

Fig. 1.2 An architecture representation of node clusters in a typical Hadoop big data processing system and the proposed scheduler optimization implemented at job tracker entity

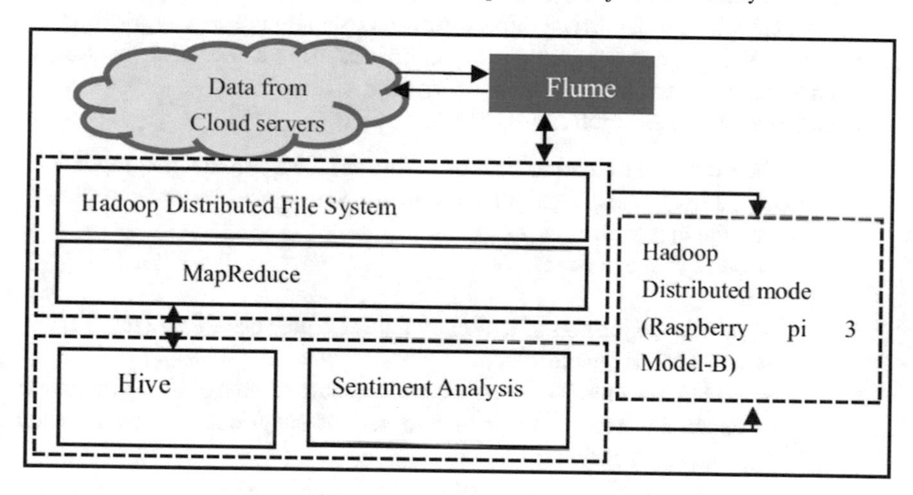

Fig. 1.3 Proposed scheduler

harnessed by Job Tracker based on the temporal demand of the job at hand and previous processing statistics of worker nodes. This results in significantly higher cluster utilization while still providing assured and reliable outputs for Hadoop workloads.

In the pseudo code, we shall define two set of Job classes, one containing the original workload allocations to data nodes and the other containing revised workload allocations to data nodes based on priority tweaks. The initial if clause of the pseudo code specifies the job request to the job tracker from the task tracker of data node which is the default setup. The execution time of the previous job is stored for prioritizing subsequent jobs. Our optimization tweak can be seen in the later if clause of the pseudo code with regard to a heart beat acknowledgement request combined with the previously stored execution time. This combined pseudo code extension to the default Hadoop setup ensures performance improvement with respect to overall execution time of processing large data sets.

1.2.2 Algorithm

1. $JobClasses1 = \{C_1, \ldots, C_n\}$, where C_i represents a class of jobs whose minimum share > 0, and share similar arrival rates, estimated execution rates and priorities
2. JobClasses2 $= \{C'_1, \ldots, C'_n\}$, where C'_i represents a category of all the jobs in the system with similar arrival rates, estimated execution rates and priorities.
3. Upon receiving a message (either a "job arrival" from a user or a "heartbeat" from a free resource) from the Hadoop system:
4. If incoming message = job (say J) arrival
   a. Get the estimated execution time of J from the Task Scheduler component.
   b. Check if J fits in the current job classification
      i. If J fits in any class (say C_i)
         - add J to the queue of C_i
      ii. Else
         - update the job classification by adding a new class for J (say C_j)
         - add J to the Queue of C_j
         - find a matching of classes and resources using an optimization approach, which results in two sets of suggested classes for each resource as follows:
           - $SC_R = \{C_1, \ldots, C_n\}, where\ C_i \in JobClasses1$
           - $SC'_R = \{C'_1, \ldots, C'_n\}, where\ C'_i \in JobClasses2$
      iii. If incoming message = heartbeat from a free resource (say R)
         - For each free slot in the resource R (say S_R)
           - Look for a job where $job.user.min.share - job.user.currentshare > 0$ and $job.class \in SC_R\ and\ ((job.user.min.share - job.user.currentshare) * priority)$ is maximum

– If no job is found, look for a job where $job.class \in SC'_R$ and $\frac{job.user.currentshare}{priority}$ is minimum

- If a job is found $(say\, J_{selected})$, send S_R and $J_{selected}$ to the Task scheduler. The Task scheduler chooses a task from $J_{selected}$ and assigns it to S_R
- If no job is found, S_R will leave unassigned until the next heartbeat message.

1.3 Results and Discussion

We design the implementation on a standard machine with the configurations Intel (R) Pentium (R) D CPU 2.8 GHz, 2 GB RAM, 3.2 bit Windows 7 operating system. The cluster and job configuration applied in our current weather temperature conversion data set are as follows. The following name node HDFS statistics snapshot validates our claim of using the 2.4 GB temperature conversion dataset (Fig. 1.4 and Table 1.1).

We have conducted rigorous evaluation on varying sizes of data sets across the native scheduler and optimized scheduler; the results show that the optimized job scheduler performs more efficiently than the Hadoop native scheduler. We also analyzed several factors that may affect job performance, including differences between CPU bound and I/O bound workloads, relationship between task and job speedup and the efficiency of sorting optimization and trade-offs but the considerable and noticeable difference is observed in the processing time. A graphical plotter describes the

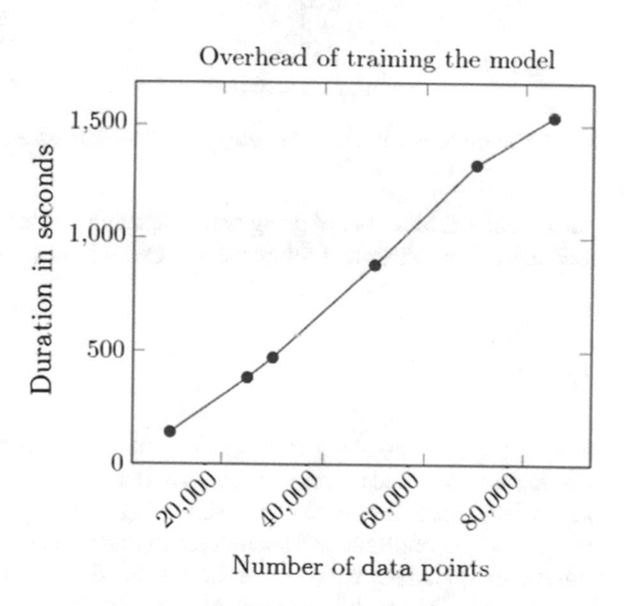

Fig. 1.4 Name node big data statistics

Table 1.1 Hadoop configuration table

Job	Temperature conversion
Compression	Disabled
Dfs. block size	128 MB
Dataset size	2.4 GB
Data nodes cluster size	3
Hadoop version	1.0.3
Slots	Map, Reduce
Virtual machine instance	1
Virtual machine tool	VMWare Cloudera-CentOS

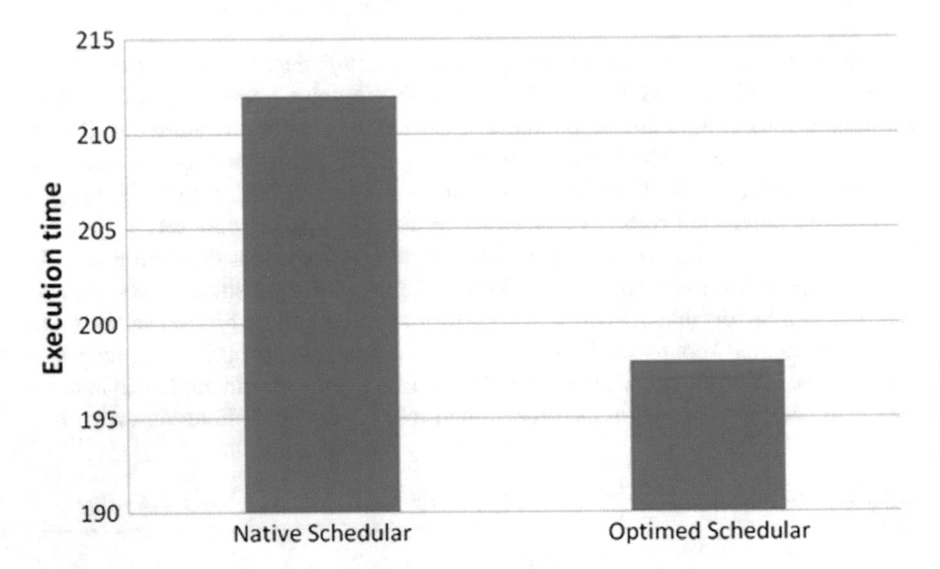

Fig. 1.5 Performance evaluation between native scheduler based and optimized scheduler based

performance differences of using native scheduler based approach and an optimized load aware job scheduler with respect to execution time (Fig. 1.5).

1.4 Conclusion

In this study, we investigated the performance bottle necks in Hadoop big data processing and identified a potential scope for improvement area with respect to scheduling at job tracker instance. For streamlining data segmentation, accumulation and assignment in MapReduce framework and to minimize system traffic and round trip times during job allocations for enormous big data applications utilizing processing nodes processing capabilities using an adaptive scheduler. We proposed an optimized two way job scheduler that considers each job's execution time of a data node and based on that decides the size of data that needs to be assigned for processing of same data node. To manage the substantial scale of big data sets of around 2.4 GB, we deployed a dispersed calculation to take care of job scheduling through pseudo distribution set up of Hadoop architecture. A possible future work might be to consider dynamic expanding big data than static fixed size data sets for processing. That could be really helpful for more real time needs of various organizations. The obtained results show that our proposition can adequately decrease system execution time under different configurations settings especially with respect to size of the data.

References

1. Samariya, D., Matariya, A., Raval, D., Babu, L.D., Raj, E.D., Vekariya, B.: A hybrid Approach for big data Analysis of cricket fan Sentiments in twitter. Paper Presented at the Proceedings of International Conference on ICT for Sustainable Development (2016)
2. Rasooli, A., Down, D.G.: A hybrid scheduling approach for scalable heterogeneous Hadoop systems. In: Proceedings of SC Companion High Performance Computing, Networking Storage and Analysis (SCC), Nov 2012, pp. 1284–1291
3. Salloum, S.A., Al-Emran, M., Monem, A.A., Shaalan, K.: A survey of text mining in social media: Facebook and twitter perspectives. Adv. Sci. Technol. Eng. Syst. J. **2**(1), 127–133 (2017)
4. Balaji, S., Golden Julie, E., Harold Robinson, Y.: Development of fuzzy based energy efficient cluster routing protocol to increase the lifetime of wireless sensor networks, Mob. Networks Appl. **24**(2), 394–406 (2019)
5. Wang, G., Gunasekaran, A., Ngai, E.W., Papadopoulos, T.: Big data analytics in logistics and supply chain management: certain investigations for research and applications. Int. J. Prod. Econ. **176**, 98–110 (2016)
6. Huang, C.-M., Shao, C.-H., Xu, S.-Z., Zhou, H.: The social Internet of Thing (S-IOT)-based mobile group handoff architecture and schemes for proximity service. IEEE Trans. Emerg. Top. Comput. **5**(3), 425–437 (2017)
7. Ahmed, S.T.: A study on multi objective clustering techniques for medical datasets" In proceedings of International Conference on Intelligent Computing and Control Systems, pp. 174–177. IEEE (2017)
8. Harold Robinson, Y., Balaji, S., Golden Julie, E.: FPSOEE: fuzzy-enabled particle swarm optimization-based energy-efficient algorithm in mobile ad-hoc networks. J. Intel. Fuzzy Syst. **36**(4), 3541–3553 (2019)
9. Eessaar, E., Saal, E.: Evaluation of different designs to represent missing information in SQL databases. In: Elleithy, K., Sobh, T. (eds.) Innovations and Advances in Computer, Information, Systems Sciences, and Engineering, pp. 173–187. Springer, New York (2013). https://doi.org/10.1007/978-1-4614-3535-8 14
10. Chang, V., Ramachandran, M.: Towards achieving big data security with the cloud computing adoption framework. IEEE Trans. Serv. Comput. **9**(1), 138–151 (2016)
11. Reddick, C.G., Chatfield, A.T., Ojo, A.: A social media text analytics framework for double-loop learning for citizen-centric public services: a case study of a local government Facebook use. Gov. Inf. Quart. **34**(1), 110–125 (2017)
12. Mittal, H., Saraswat, M.: An optimum multi-level image thresholding segmentation using non-local means 2D histogram and exponential Kbest gravitational search algorithm. Eng. Appl. Artif. Intell. **71**, 226–235 (2018)
13. Gong, B., Veeravalli, B., Feng, D., Zeng, L., Wei, Q.: CDRM: a cost-effective dynamic replication management scheme for cloud storage cluster. In: 2010 IEEE International Conference on Cluster Computing, Sept 2010, pp. 188–196 (2010)

14. Balaji, S., Golden Julie, E., Harold Robinson, Y., Raghvendra, K., Thong, P.H., Le Hoang, S.: Design of a security-aware routing scheme in mobile ad-hoc network using repeated game model. Comput. Stan. Interfaces **66** (2019)
15. Hoeber, O., Hoeber, L., El Meseery, M., Odoh, K., Gopi, R.: Visual Twitter Analytics (Vista) Temporarily changing sentiment and the discovery of emergent themes within sport event tweets. Online Inf. Rev. **40**(1), 25–41 (2016)
16. Poria, S., Cambria, E., Howard, N., Huang, G.-B., Hussain, A.: Fusing audio, visual and textual clues for sentiment analysis from multimodal content. Neurocomputing **174**, 50–59 (2016)
17. Ashish, T., Kapil, S., Manju, B.: Parallel bat algorithm-based clustering using MapReduce. In: Networking Communication and Data Knowledge Engineering, pp. 73–82. Springer, Singapore (2018)
18. Kangin, D., Angelov, P., Iglesias, J.A., Sanchis, A.: Evolving classifier for big data. Procedia Comput. Sci. **53**, 9–18 (2015). In: Conference on Big Data 2015 Program San Francisco, CA, USA, 8–10 Aug 2015
19. Harold Robinson, Y., Rajaram, M.: A memory aided broadcast mechanism with fuzzy classification on a device-to-device mobile ad hoc network. **90**(2), 769–791 (2016)
20. Oussous, A., Benjelloun, F.Z., Ait Lahcen, A., Belfkih, S.: Big data technologies: a survey. J. King Saud Univ. Comput. Inf. Sci. **30**, 431–448 (2017)
21. Harold Robinson, Y., Rajaram, M.: Energy-aware multipath routing scheme based on particle swarm optimization in mobile ad hoc networks. Sci. World J. 1–9 (2015)

Chapter 2
Big Data and Machine Learning for Evaluating Machine Translation

Rashmi Agrawal and Simran Kaur Jolly

Abstract Human Evaluation of Machine Translation is the most important aspect of improving accuracy of translation output which can be used for text categorization ahead. In this article we describe approach of text classification based on parallel corpora and natural language processing techniques. A text classifier is built on multilingual texts by translating different features of the model using the Expectation Maximization Algorithm. Cross-lingual text classification is the process of classifying text into different languages during translation by using training data. The main idea underlying this mechanism is using training data from parallel corpus and applying classification algorithms for reducing the distortion and alignment errors in Machine translation. In this chapter a Classification Model is trained which directs source language to target language on the basis of translation knowledge and parameters defined. The Algorithm adopted here is Expectation Maximization Algorithm which removes ambiguity in parallel corpora by aligning source sentence to target sentence. It considers possible translations from source to target language and selects the one that fits the model on the basis of BLEU (bilingual evaluation understudy) score. The only requirement of this learning is unlabelled data in the target language. The algorithm can be evaluated accurately by running a separate classifier on different parallel corpora. We use Monolingual Corpora and Machine Translation in our study to see the effect of both the models on our parallel corpora.

Keywords Corpus · Classifier · Machine translation · Categorization · Expectation maximization

R. Agrawal (✉) · S. K. Jolly (✉)
Faculty of Computer Applications, MRIIRS, Faridabad, Haryana, India
e-mail: drrashmiagrawal78@gmail.com

S. K. Jolly
e-mail: Simrankaur1611@yahoo.com

V. E. Balas et al. (eds.), *Internet of Things and Big Data Applications*, Intelligent Systems Reference Library 180, https://doi.org/10.1007/978-3-030-39119-5_2

2.1 Introduction

Due to accelerated growth of multilingual documents on the internet, various cross language text classification techniques are being effective and important. Large number of text preprocessing tools is being developed for corpus development, machine translation and word sense disambiguation [1] and others. The Cross lingual text classification task addresses the issue of classifiers for classifying the text data in different languages to avoid time complexity. It assigns semantic classes to the documents in target language by training the data in source language. Initially text classification was done on monolingual data by translating the training data to test data into the same language format. The drawbacks of monolingual classification systems is that it only generates best translation out of the given corpus domain. Training the data again and again may diminish the classification accuracy. Domain adaptation is another drawback due to difference in training and test datasets, which causes monolingual data to not perform well. The approaches for multilingual text classification is classified into knowledge based approaches and empirical approaches. Knowledge based approaches are rule based approaches having low accuracy as it is difficult tweaking large lexicons foe small tasks like word sense disambiguation, extending lexicon with translations. The empirical based approaches rely on statistical methods, as they are trained using already aligned corpora.

2.1.1 Types of Corpora

The corpora taken in consideration for training of the textual data are of two types:

a. **Parallel Corpora**: Parallel corpora are texts in two different languages, where one of the source text is source language and target text is the target language. These texts are bit-text which may be unidirectional, bidirectional or multidirectional [2].
b. **Comparable Corpora**: Comparable corpora are collections of texts in different languages belonging to same domain. Parallel corpora can be comparable in nature but comparable corpora are multilingual in nature. Hence the comparable corpora are bidirectional in nature.

2.1.2 Automatic Classification

Text classification of documents is an automatic process that includes classifier, universal encoder and sample of documents. The corpus that we are using in text classification is comparable corpora. Language classification consists of two steps (i) Universal Encoder (ii) Classifier (Fig. 2.1).

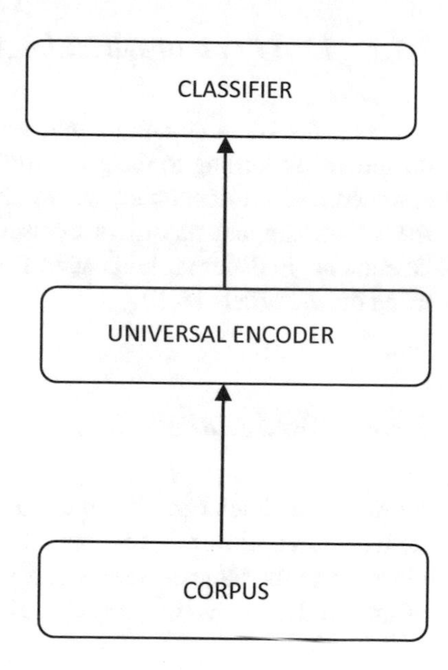

Fig. 2.1 Automatic classification

a. **Universal Encoder**: The Universal Encoder in the classification algorithm trains the large labeled texts in different languages. The Universal Encoder uses a comparable corpus, which uses a zero shot classification where the entire text in target language is encoded, trained to predict the class of the document. This also helps in removing ambiguity problems in the corpus. In this process the text in the source language is encoded into some encoded form, which is then compared with the target language text and then the classifier is applied over the comparable corpus.
b. **Classifier**: The output of the universal encoder is trained on the classifier. The model is trained on sentences ranging from 5 k to hundred thousand sentences. It is a model based on machine translation that encodes the sentences and then train the model to classify it according to the classifier.

2.2 Approaches for Classification

A comparable corpus can be easily built as compared to our parallel corpora. Wikipedia is example of comparable corpora. In the given classification task [3] we use news articles from different domains like sports or finance from Wikipedia dumps. The following approaches are being used for document classification.

2.2.1 LDA (Latent Dirichlet Allocation)

This is a statistical technique where which uses Bayes classifier for arranging the documents according to the given topic. In this technique probability distribution is assigned to each document, where distribution of words is formed for each document. Cross lingual classifiers are used for text classification and text categorization. Documents in different languages are classified according to semantic distribution using the classifiers [3–5].

2.2.2 Word Embeddings

Word Embeddings can be trained using lexicons and unlabelled text. For e.g. the sentence blue flower can be translated to blaue flower where the words blue and blaue are of the same context according to context bag of words. Hence the meaning of the word is known by its surrounding word.

2.2.3 Structural Correspondence Learning

It is equivalent to machine learning process where the pivot words are used in both the source and target language. Pivot words are union of source and target language. For each pivot a classifier is trained to predict the words in the text. It is just trained using the corpus in source language, pivots and target language [6].

2.3 Automatic Metrics

2.3.1 Word Error Rate

This is one of the most important techniques in machine translation. It is the sum of substitution, insertion and deletion divided by total number reference sentences. It uses Levenshtein distance as the metric. This metric is used for sentence alignment in machine translation. It assigns either 0 or 1 value to the aligned sentence.

2.3.2 BLEU

It is precision oriented machine translation evaluation technique that counts number of sentences that are correct in the corpus with respect to the reference translation [7].

$$\text{Precision} = \text{count clip(n gram)}/\text{count(n gram)}$$

The shortcoming of the metric is it cannot be used for synonym match and inability to detect word orders.

2.3.3 Meteor

It is recall oriented metric that computes both precision and recall for the sentence or document by computing the harmonic mean. In-order to align the two sentences these metrics are used:

a. **Exact Matching**: The strings that are perfectly aligned by ratio 1:1.
b. **Stem Matching**: In this the words are stemmed to their root form and then aligned.
c. **Synonym Matching**: In this technique words which are similar to each other are aligned by using a dictionary and lexicon. Precision in this metric is defined by number of words matching divided by total number of words in the corpus. Recall in this metric is defined by number of words matching divided by total number of words in the reference.

2.3.4 Translation Edit Rate

The translation edit rate captures the movement of the phrases in source and target language. **It captures the maximum sequence of the sentences that are matched using the formula** size (M) $= \sum_{r£M}$ length $(r)^2$. The translation edit rate addresses the phrase reordering by counting number of shifts in the sentences. It uses a greedy search method for modeling the quality of translation. The constraints used by translation edit rate are as follows:

a. Shifts are selected by greedy approach which select word error rate having minimum value.
b. Sequence of the words shifted in hypothesis must be equal to sequence of words shifted in reference.
c. In-order to correctly match words, it should have at least one error.
d. The matched words in the reference translation must have at least one error.

2.4 Related Work

Automatic systems that can classify documents quickly and precisely are useful for a wide range of practical applications. For example, organizations may be interested in using sentiment analysis of opinion posts such as tweets that mention their products and services. By classifying the sentiment of each post (e.g. positive, neutral, or negative); the organization can for example learn which parts of a product should be improved. Creating a suitable, large labeled dataset for training a classification model requires a lot of effort and available public datasets are typically only available in the most common languages. In order to train a classification model for a languages Lt without a suitable text classification dataset there are two options: The first option is of course to create a new labeled dataset from scratch, and the second option is to use the label information in existing labeled datasets in a language Ls and then transfer this label information to Lt. The first option usually requires a great amount of work and is typically not a viable solution.

The second option is called cross-language text classification (CLTC) [5]. In this article we present a method for performing CLTC by means of a universal encoder. The method consists of two steps. In the first step, a universal encoder is trained to give similar representations to texts that describe the same topic, even if the texts are in different languages. In the second step, a classification module uses the language-independent representations from the universal encoder as inputs and is trained to predict which category each document belongs to. Compared to previous work, this method has several advantages: (1) the universal encoder can be trained using just a comparable corpus. A comparable corpus is a corpus where the multilingual versions of a document are not necessarily translations of each other, but merely about the same topic. (2) It enables zero-shot classification. I.e. if we have a comparable corpus in French-Spanish, we can build a classifier for Spanish by using a labeled dataset in English. (3) The universal encoder does not rely on single word translations, but rather on encoding entire contexts. This can help alleviate ambiguity problems caused by polysemy. The input language does not have to be specified at test time.

The method presented in this chapter is conceptually similar in spirit to Google's zero-shot machine translation model [8], which is used in the Google translate API. That model also uses a shared vocabulary and a language independent encoder. It does, however, require a large corpus of aligned sentences for training. Additionally, translating a text is a much harder problem than merely extracting discriminative features since it requires encoding of e.g. syntactic information that is not necessary for text classification. Therefore such a model is much more complex than it needs to be, and a more parsimonious model is therefore preferable.

Multilingual word embeddings can also be trained using only a dictionary and unlabeled text [9]. This can be done by switching words with the same meaning in different languages. E.g. the sentence The red hand could be modified to The rojo hand using a dictionary.

These artificially modified sentences can then be used to train multilingual word embeddings by using e.g. a CBOW [10] model. There are some challenges to using this kind of model, however. For example, the method relies on the fact that the meaning of a word is determined by its context (the distributional hypothesis). But in the example above readers familiar with Spanish grammar rules will know that rojo and hand would not belong to the same context (but roja and hand would).

2.5 Text Alignments

Text alignment is the quintessential process required for text classification. In-order to make text classification accurate a written corpus is needed. Written corpus is collection of parallel text for developing natural language processing tasks like machine translation. The corpus can be either annotated or un-annotated. The annotated corpus has labeled sentences having specific attributes like parts of speech tagger, topics, domains and mood of the text. For example, the labeled attributes in a corpus for the word "roses" could be *noun*, *plural*, etc. Linguistic information that could be labeled would be its *lemma2*, the correct sense of the word according to a specific dictionary, etc.; and, in other languages like Spanish, labels like *feminine noun* or *masculine noun* could be added. The corpus below shows a sample of the SemEval-2015 task 13 corpus which consists in four documents taken of European Medicines Agency documents, the KDE manual corpus and the EU bookshop corpus (Figs. 2.2 and 2.3).

```
<? xml version="1.0" encoding="UTF-8" ?>
<corpus lang="english">
<text id="t001">
<sentence id="t001.s001">
<wf id="t001.s001.w001" pos="X">This</wf>
<wf id="t001.s001.w002" lemma="document" pos="N">document</wf>
<wf id="t001.s001.w003" lemma="be" pos="V">is</wf>
<wf id="t001.s001.w004" pos="X">a</wf>
<wf id="t001.s001.w005" lemma="summary" pos="N">summary</wf>
<wf id="t001.s001.w006" pos="X">of</wf>
<wf id="t001.s001.w007" pos="X">the</wf>
<wf id="t001.s001.w008" lemma="european" pos="J">European</wf>
<wf id="t001.s001.w009" lemma="public" pos="J">Public</wf>
<wf id="t001.s001.w010" lemma="assessment" pos="N">Assessment</wf>
<wf id="t001.s001.w011" lemma="report" pos="N">Report</wf>
```

Fig. 2.2 Corpus structure

			Target text		
			t1	*t2*	*t3*
			Mi nombre es Guiller-mo y puedo leer en inglés y español	Sin embargo quiero aprender otros idiomas	Esto es un texto en inglés
Source text	*s 1*	My name is William			
	s 2	I can read English and Spanish texts			
	s 3	This is an English text			

1 to 1 alignment
1 to 2 alignment
2 to 1 alignment
2 to 2 alignment

Fig. 2.3 Sentence alignment template

2.5.1 Algorithm

The vector model is a multi dimensional space where all the words in the text are represented in form of vector. The dataset for evaluation is divided into two parts: encoder and classification.

The basic assumption of classification algorithm is that source and the target language convey same meaning in different ways. We use this information to train the classifiers. The learning paradigm relies on two parameters:

a. A monolingual misclassification cost for each classifier in each language/view.
b. A disagreement cost to constrain decisions to be similar in both languages.

The vector model assumed has two functions f1 and f2 which have real valued functions h1 and h2 which are controlled using sign function. Our framework works by optimizing the classifier h from one view (h = h, ∈ {1, 2}), view while holding the classifier from the other view (h∗ = h3−) fixed. This is done by minimizing a monolingual classification loss in that view and regularizing it by a divergence term which constrains the output of the trained classifier to be similar to that of the classifier previously learned in the other view. Without losing the originality, we describe the stage where it optimizes functions h from one view, while leaving the function from the other view, h∗, fixed. The following assumptions when followed lead us to this equation:

$$L(h, S, h*, S*, \lambda) = C(h, S) + \lambda d(h, S, h*, S*).$$

C(h, S) = monolingual cost of training h on the set S
λd(h, S, h∗, S∗) = divergence between two documents

for monolingual cost the misclassification error is computed as $= 1/m \sum_{i=1}^{m}$ yi h(xi) < 0.

Algorithm

Input: Two labeled sets S1 and S2;
A discount factor λ.
Initialize: t ← 1;
h(0) 1 def = argminh C(h,S1);
h(0) 2 def = argminh C(h,S2);
repeat
Learn h(t) 1 = argminh L(h,S1,h(t − 1) 2,S2,λ);
Learn h(t) 2 = argminh L(h,S2,h(t) 1,S1,λ);
t ← t + 1;
until Convergence of ▲(h(t) 1,S1,h(t) 2,S2,λ) (eq. 4) to a local minimum;
Output: f1 = sign(h(t) 1) and f2 = sign(h(t) 2)
▲(h1,S1,h2,S2,λ) = C(h1,S1) + C(h2,S2) (misclassification) +λD(h1,S1,h2,S2) (disagreement).

2.5.2 *Text Classification Using Supervised Learning*

Support Vector Machines and Naïve Bayes Classifier is used for classifying the documents into different categories for text classification, machine translation and text categorization. The dataset that we will be using is the Amazon review dataset.

1. Adding the Libraries: The first step for supervised text classification is adding the defined libraries into your python editor (Fig. 2.4).
2. Uploading the Corpus: The corpus is uploaded using the pandas library in the python. It is uploaded in csv format.
3. Data Pre-Processing: This is the most quintessential step of classification algorithm. It converts the raw text into an understandable format for Natural Language Processing Models like text classification and machine translation.

```
import pandas as pd
import numpy as np
from nltk.tokenize import word_tokenize
from nltk import pos_tag
from nltk.corpus import stopwords
from nltk.stem import WordNetLemmatizer
from sklearn.preprocessing import LabelEncoder
from collections import defaultdict
from nltk.corpus import wordnet as wn
from sklearn.feature_extraction.text import TfidfVectorizer
from sklearn import model_selection, naive_bayes, svm
from sklearn.metrics import accuracy_score
```

Fig. 2.4 Adding libraries

The steps involved in Data Pre-processing are:

a. Tokenization: It is the process of breaking sequence of sentences into words, phrases and segments using word_tokenize and sent_tokenize library.
b. Word Stemming: It is the process of reducing the word into its inflectional form. The words when reduced to it stemmed form is easier to implement (Fig. 2.5).

4. Division of the Corpus: The corpus is converted into two parts, training and testing data. The train and test data is derived from sklearn library (Fig. 2.6).
5. Word Vectorization: It is the process of converting text into a vector format. The words in the corpus are converted into tf and idf format. It basically builds vocabulary of the words through the corpus having some frequency.

```
    Final_words = []
    # Initializing WordNetLemmatizer()
    word_Lemmatized = WordNetLemmatizer()
    # pos_tag function below will provide the 'tag' i.e if the word
is Noun(N) or Verb(V) or something else.
    for word, tag in pos_tag(entry):
        # Below condition is to check for Stop words and consider
only alphabets
        if word not in stopwords.words('english') and word.isalpha():
            word_Final =
word_Lemmatized.lemmatize(word,tag_map[tag[0]])
            Final_words.append(word_Final)
    # The final processed set of words for each iteration will be
stored in 'text_final'
    Corpus.loc[index,'text_final'] = str(Final_words)
```

Raw Text	Pre-processed Text
Stuning even for the non-gamer: This sound track was beautiful! It paints the senery in your mind so well I would recomend it even to people who hate video game music! I have played the game Chrono Cross but out of all of the games I have ever played it has the best music! It backs away from crude keyboarding and takes a fresher step with grate guitars and soulful orchestras. It would impress anyone who cares to listen! ^_^	['stun', 'even', 'sound', 'track', 'beautiful', 'paint', 'senery', 'mind', 'well', 'would', 'recomend', 'even', 'people', 'hate', 'video', 'game', 'music', 'play', 'game', 'chrono', 'cross', 'game', 'ever', 'play', 'best', 'music', 'back', 'away', 'crude', 'keyboarding', 'take', 'fresh', 'step', 'grate', 'guitar', 'soulful', 'orchestra', 'would', 'impress', 'anyone', 'care', 'listen']

Fig. 2.5 Text pre-processing

```
Train_X, Test_X, Train_Y, Test_Y =
model_selection.train_test_split(Corpus['text_final'],Corpus['label']
,test_size=0.3)
```

Train_X → Training Data Predictors
Train_Y → Training Data Target
Test_X → Test Data Predictors
Test_Y → Test Data Target

Fig. 2.6 Splitting the train and test data

```
# Classifier - Algorithm - SVM
# fit the training dataset on the classifier
SVM = svm.SVC(C=1.0, kernel='linear', degree=3, gamma='auto')
SVM.fit(Train_X_Tfidf,Train_Y)

# predict the labels on validation dataset
predictions_SVM = SVM.predict(Test_X_Tfidf)

# Use accuracy_score function to get the accuracy
print("SVM Accuracy Score -> ",accuracy_score(predictions_SVM,
Test_Y)*100)
```

Fig. 2.7 Generating accuracy

a. Term Frequency: This summarizes how many times word appear in a document.
b. Inverse Document Frequency: This scales words that appear frequently in documents.

6. Machine Learning: Support Vector Machines is a machine learning algorithm which divides the entire training dataset into two parts using a hyper-plane. The accuracy generated by the svm classifier is 84% (Fig. 2.7).

2.5.3 Text Classification Using Unsupervised Algorithm

This is a problem where text is classified into clusters. The clusters are similar to each other by grouping set of unlabelled texts into one cluster. K-means algorithm [11] is applied on the corpus by assuming k seeds and then dividing them into k clusters.

1. Corpus: The first step is building the corpus. So we build corpus based on cricket and travelling topic and assuming k = 2 (Fig. 2.8).

In the given corpus above we classify it according to different classes. So we extract those classes form the corpus using un-supervised k-means clustering algorithm.

2. Applying TF_IDF and K-Means Algorithm

The following libraries are imported first in-order to extract the features from the corpus.

```
document = ["This is the most beautiful place in the world.", "This
man has more skills to show in cricket than any other game.", "Hi
there! how was your ladakh trip last month?", "There was a player
who had scored 200+ runs in single cricket innings in his career.",
"I have got the opportunity to travel to Paris next year for my
internship.", "May be he is better than you in batting but you are
much better than him in bowling.", "That was really a great day for
me when I was there at Lavasa for the whole night.", "That's exactly
I wanted to become, a highest ratting batsmen ever with top
scores.", "Does it really matter wether you go to Thailand or Goa,
its just you have spend your holidays.", "Why don't you go to
Switzerland next year for your 25th Wedding anniversary?", "Travel
is fatal to prejudice, bigotry, and narrow mindedness., and many of
our people need it sorely on these accounts.", "Stop worrying about
the potholes in the road and enjoy the journey.", "No cricket team
in the world depends on one or two players. The team always plays to
win.", "Cricket is a team game. If you want fame for yourself, go
play an individual game.", "Because in the end, you won't remember
the time you spent working in the office or mowing your lawn. Climb
that goddamn mountain.", "Isn't cricket supposed to be a team sport?
I feel people should decide first whether cricket is a team game or
an individual sport."]
```

Fig. 2.8 Corpus

```
from sklearn.feature_extraction.text import TfidfVectorizer
from sklearn.cluster import KMeans
import numpy as np
import pandas as pd
```

Tf-Idf vectorizer is an important method of converting textual information into binary vector form. It basically assigns vector number to a word on basis of its occurrence (Fig. 2.9).

K-means is an unsupervised approach followed by assuming certain number of fixed clusters (k = 2) and then calculating the distance of the word from the centroids. The centroids are placed far away from each other and then words are grouped into clusters accordingly (Figs. 2.10 and 2.11).

```
vectorizer = TfidfVectorizer(stop_words='english')
X = vectorizer.fit_transform(document)
```

Fig. 2.9 Count vectorizer

```
true_k = 2
model = KMeans(n_clusters=true_k, init='k-means++', max_iter=100,
n_init=1)
model.fit(X)
```

We will get the below output:

```
Out[5]:
KMeans(algorithm='auto', copy_x=True, init='k-means++',
max_iter=100,
    n_clusters=2, n_init=1, n_jobs=1, precompute_distances='auto',
    random_state=None, tol=0.0001, verbose=0)
```

Fig. 2.10 K-means algorithm

	0	1	2	3	4
0	7	24	38	44	29
1	12	77	89	91	62

Centroids

terms - List (92 elements)

Index	Type	Size	Value
0	str	1	200
1	str	1	25th
2	str	1	accounts
3	str	1	anniversary
4	str	1	batsmen
5	str	1	batting
6	str	1	beautiful
7	str	1	better
8	str	1	bigotry

Fig. 2.11 Generating centroids

```
print("\n")
print("Prediction")
X = vectorizer.transform(["Nothing is easy in cricket. Maybe when
you watch it on TV, it looks easy. But it is not. You have to use
your brain and time the ball."])
predicted = model.predict(X)
print(predicted)
```

This will give the following output:

```
Prediction
[1]
```

Fig. 2.12 Classification of sentences

After getting the terms out of the corpus, we fit the terms into the clusters. After fitting the words into clusters we predict the sentences in the test corpus we try to predict the clusters of the sentences (Fig. 2.12).

Hence we try to classify or unlabelled data using the k-means classifier algorithm, the results are not accurate always and depends on the quality of corpus we are taking as input in the model.

2.5.4 *Text Classification Using Semisupervised Learning Using EM Algorithm*

In this approach Expectation Maximization Algorithm is used to classify text by extracting features from the translational probabilities of labeled text in source language and unlabelled text in target language. Hence the objective of text classification is to reduce errors due to word disambiguation and classify without erroneous words. The text classification works as follows:

a. Extracting Features: in order to extract features from set of documents D in the target language we have to generate probability of generating any document d from mixture of different classes' c. According to bag of word models in natural language processing: $P(d) = \sum c\ P(c) \sum d' \prod P(wi|w'i, c)P(w'i|c)$

In the above equation wi is the ith word of the document d with l words. The prior probability P(c) and the probability of the source language word w′ given class c are estimated using the labeled training data in the source language, so we use them as known parameters. P(wi|w′i, c) is the probability of translating the word w′ i in the source language to the word wi in the target language given class c, and these are the parameters we want to learn from the corpus in the target language. So the feature that is learnt is P(wi|w′i, c) from the equation.

Algorithm: Semi-supervised learning for crosslingual text classification

Ls ← labeled data in the source language Ut ← unlabeled data in the target language

1: Cs = train(Ls)
2: Ct = translate(Cs)
3: repeat
4: Label(U, Ct)
5: L ← select(confidence(U, Ct))
6: Ct ← train(L)
7: until stopping criterion is met
8: return Ct

In order to evaluate the technique we take dataset from English, French and Chinese in 5 different domains.

Figure 2.13 shows distribution of the documents in different domains. In order to build the classifier from English-Chinese and English-French language pair we take English as source language and Chinese and French as the target language. To transfer the model from source to target language the features are converted to bag of words. The three methods EQUAL, UNIGRAM and EM are applied to the given corpus (Table 2.1).

In the equal method different translation probabilities are assigned to the word in the source language and then the word in the target language is randomly picked. This is considered the baseline method having lowest frequency as it may result in ambiguous word pair. While in Unigram method the translation probabilities of the words are calculated on the basis of their frequency. In EM method the probabilities are learned both from source and target document. We evaluated all the three models on our parallel corpus and achieved highest precision value in the EM method. Hence classifying the data with semi-supervised classifier yielded better accuracy as the model fitted the data accurately.

Fig. 2.13 Dataset

Category	English	Chinese	French
sports	23764	14674	18398
health	15627	11769	12745
business	34619	23692	28740
entertainment	26876	21470	23756
education	16488	14353	15753

Table 2.1 Comparison of precision for the methods

Method	P	R	F
EQUAL	71.1	70.6	70.8
UNIGRAM	79.5	77.8	78.6
EM	83.1	84.7	83.9

2.5.5 *Gaussian Mixture Models*

Gaussian Mixture Models are unsupervised techniques used for clustering approach over the classical nearest neighbor algorithm. The algorithm behind the Gaussian mixture model is the Expectation Maximization Algorithm. In the Expectation step the probability of the data-points is calculated and in the maximization step the log-likelihood function is maximized for each of the point in the cluster.

The whole model is implemented on the above dataset defined above in-order to cluster the corpus into two different classes (Figs. 2.14, 2.15 and 2.16).

Hence we can see in the above plots that the corpus we had was a mixed corpus having sentences of different domains, the number of clusters were known, but not labeled. Hence using the above Gaussian distribution we could compute probabilities of data points or the words. The words having the same probabilities are labeled in the same cluster or domain.

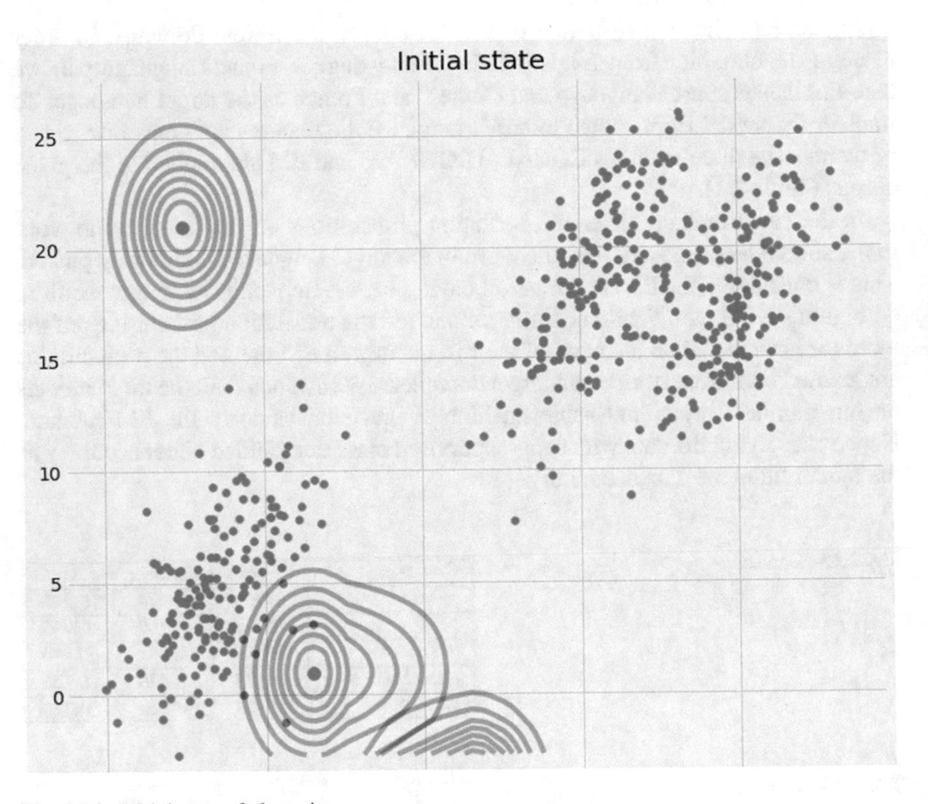

Fig. 2.14 Initial state of clustering

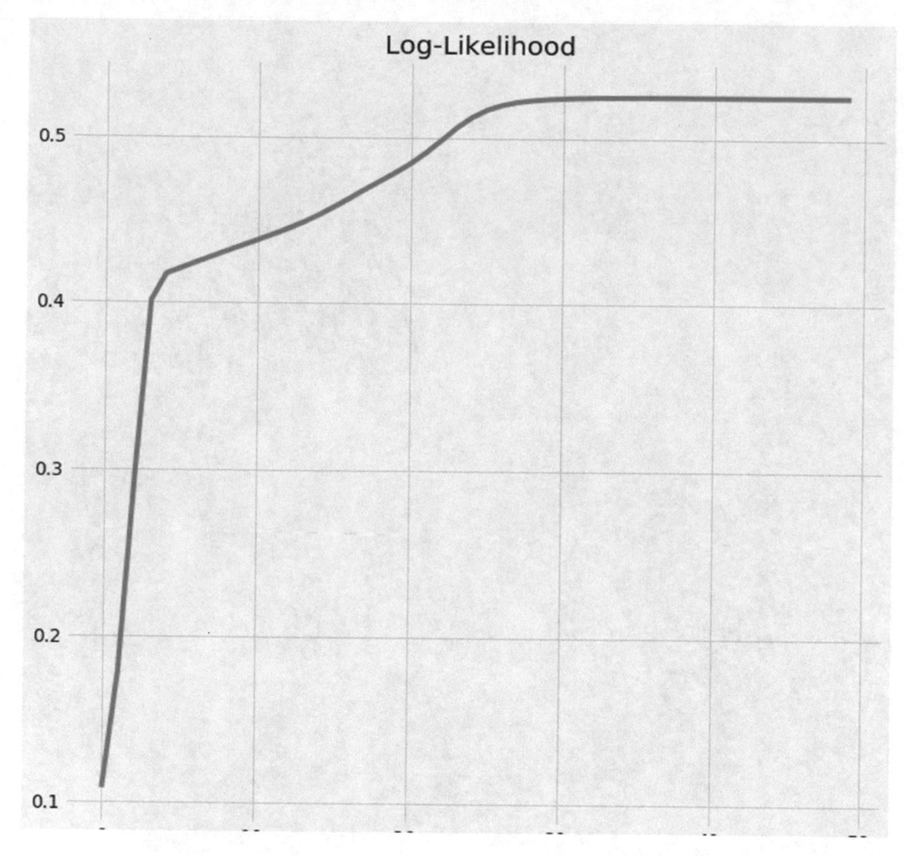

Fig. 2.15 Generating log-likelihood curve

2.5.6 *Text Classification Using doc2vec and Regression*

The goal of this technique is to classify the data using doc2vec technique and then training the data through logistic regression classifier. The dataset taken for training is consumer complaints and we have to classify them into 12 classes. We will be Gensim toolkit in natural language processing for text preprocessing and classification.

1. In the Fig. 2.17 following complaints and there classes are shown. As the dataset is very large we need to reduce the dataset and index into data-frames.
2. In the next step the data is divided into training and test dataset.

   Figure 2.18 shows the words in the training dataset having tags credit reporting.

3. Building a Vocabulary

   After training the above doc2vec model in Gensim for 30 s following results are obtained (Figs. 2.19 and 2.20).

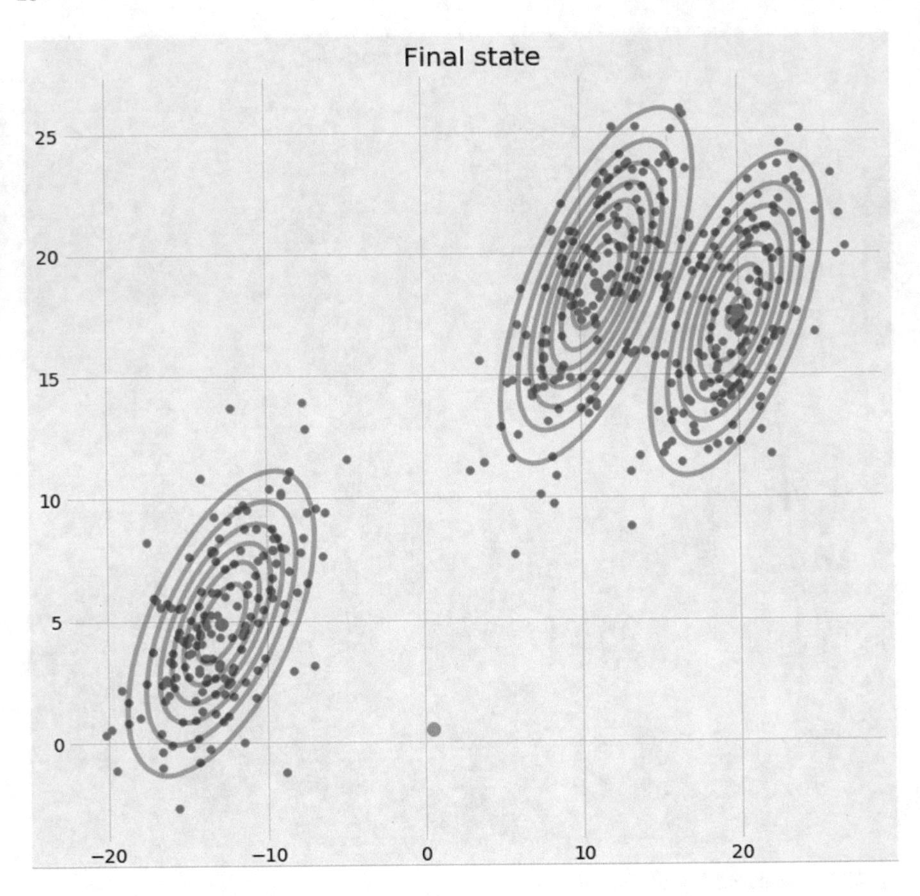

Fig. 2.16 Corpus clustering

	narrative	Product
1	When my loan was switched over to Navient i wa...	Student loan
2	I tried to sign up for a spending monitoring p...	Credit card or prepaid card
7	My mortgage is with BB & T Bank, recently I ha...	Mortgage
14	The entire lending experience with Citizens Ba...	Mortgage
15	My credit score has gone down XXXX points in t...	Credit reporting
17	I few months back I contacted XXXX in regards...	Credit reporting, credit repair services, or o...
28	I " m a victim of fraud and I have a file wit...	Credit reporting, credit repair services, or o...
30	My mortgage is owned by XXXX, we have painfull...	Mortgage
32	I have been disputing a Bankruptcy on my credi...	Credit reporting, credit repair services, or o...
54	Today I received a phone call from a number li...	Debt collection

Fig. 2.17 Classification of sentences

```
train_tagged.values[30]
```

```
TaggedDocument(words=['had', 'bankruptcy', 'years', 'ago', 'and', 'it', 'is', 'still', 'showing', 'up', 'on', 'equifa', 'whic
h', 'is', 'preventing', 'me', 'from', 'buying', 'home', 'at', 'good', 'rate', 'they', 'need', 'to', 'take', 'it', 'off', 'lik
e', 'did', 'so', 'my', 'score', 'will', 'be'], tags=['Credit reporting'])
```

Fig. 2.18 Building vocabulary

```
100%|██████████| 223102/223102 [00:00<00:00, 1596907.77it/s]
100%|██████████| 223102/223102 [00:00<00:00, 2576198.16it/s]
100%|██████████| 223102/223102 [00:00<00:00, 1961591.50it/s]
100%|██████████| 223102/223102 [00:00<00:00, 1799270.90it/s]
100%|██████████| 223102/223102 [00:00<00:00, 1216525.65it/s]
100%|██████████| 223102/223102 [00:00<00:00, 1879469.14it/s]
100%|██████████| 223102/223102 [00:00<00:00, 2159058.65it/s]
100%|██████████| 223102/223102 [00:00<00:00, 1851940.21it/s]
100%|██████████| 223102/223102 [00:00<00:00, 1868324.12it/s]
100%|██████████| 223102/223102 [00:00<00:00, 2222950.33it/s]
100%|██████████| 223102/223102 [00:00<00:00, 2035760.37it/s]
100%|██████████| 223102/223102 [00:00<00:00, 1791036.78it/s]
100%|██████████| 223102/223102 [00:00<00:00, 2039549.64it/s]
100%|██████████| 223102/223102 [00:00<00:00, 3569332.45it/s]
100%|██████████| 223102/223102 [00:00<00:00, 2855374.47it/s]
100%|██████████| 223102/223102 [00:00<00:00, 2855418.04it/s]
100%|██████████| 223102/223102 [00:00<00:00, 2379567.07it/s]
100%|██████████| 223102/223102 [00:00<00:00, 2039580.76it/s]
100%|██████████| 223102/223102 [00:00<00:00, 2322399.87it/s]
100%|██████████| 223102/223102 [00:00<00:00, 2517534.79it/s]
100%|██████████| 223102/223102 [00:00<00:00, 1888184.11it/s]
100%|██████████| 223102/223102 [00:00<00:00, 919457.47it/s]
100%|██████████| 223102/223102 [00:00<00:00, 2060277.88it/s]
100%|██████████| 223102/223102 [00:00<00:00, 1885517.07it/s]
100%|██████████| 223102/223102 [00:00<00:00, 1539867.75it/s]
100%|██████████| 223102/223102 [00:00<00:00, 2726717.42it/s]
100%|██████████| 223102/223102 [00:00<00:00, 2033057.94it/s]
Wall time: 18min 24s
```

Fig. 2.19 Model training using Gensim

The above model is trained using the logistic regression classifier. The waiting time in logistic regression classifier was much more than as compared to the bag of word models. Hence both the models are concatenated to achieve accuracy greater than the individual models. The overall accuracy of the model was around 67% by using distributed memory with averaging, where the vector size of words was increased iterating those over 30 epochs. Hence model pairing is an important technique being used to achieve high accuracy and better classification in large corpus.

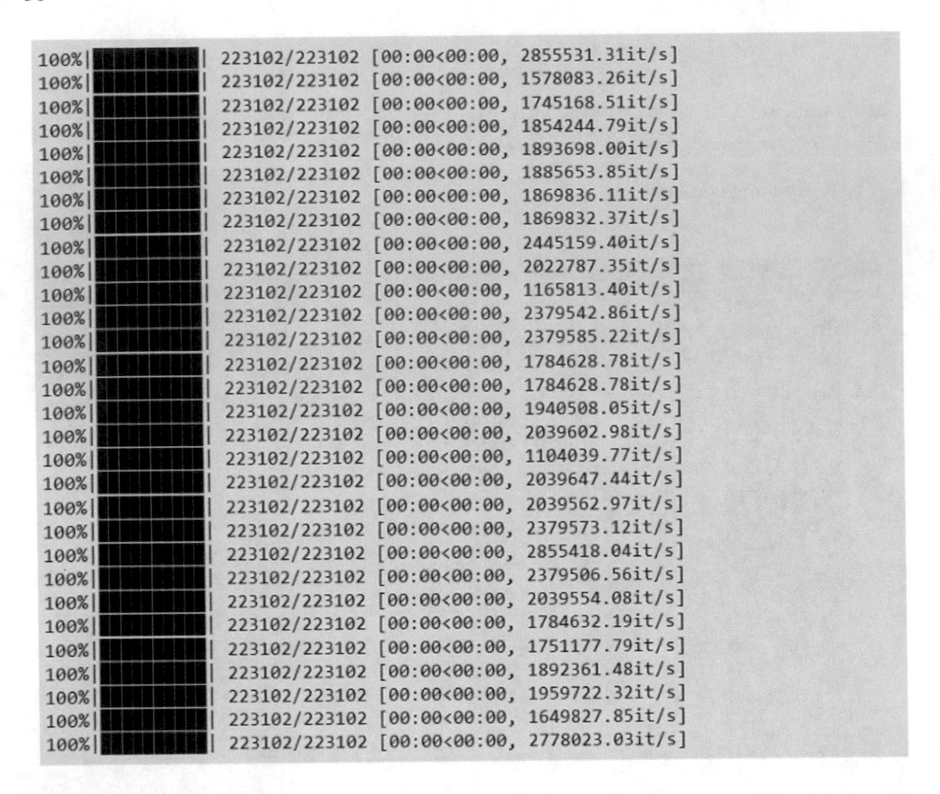

Fig. 2.20 Training the corpus

2.5.7 Text Classification Using BLEU

BLEU score is generated by comparing sentences of candidate sentences to reference translations. It is presented in NLTK library in python. The algorithm works by matching the unigrams in the sentences regardless of the word alignment in sentences. It assigns value 0 or 1 to sentence if the unigram match exists.

Reference: The NASA Opportunity rover is battling a massive dust storm on Mars.
Candidate 1: The Opportunity rover is combating a big sandstorm on Mars.
Candidate 2: A NASA rover is fighting a massive storm on Mars.

The above sentence pair consists of two translations and one reference translation. In order to compute the bleu score of the candidate translations the sentences are first tokenized and then the BLEU score of the sentences are computed.

The following metrics are used for bleu score computation.

N-Gram Precisions

Table 2.2 contains the n-gram precisions for both candidates.

Table 2.2 Metric calculation

Metric	Candidate 1	Candidate 2
precision1 (1gram)	8/11	9/11
precision2 (2gram)	4/10	5/10
precision3 (3gram)	1/9	2/9
precision4 (4gram)	0/8	1/8
Brevity-penalty	0.83	0.83
BLEU-score	0.0	0.27

- **Brevity-Penalty**
  The brevity-penalty is the same for candidate 1 and candidate 2 since both sentences consist of 11 tokens.
- **BLEU-Score**
  Note that at least one matching 4-gram is required to get a BLEU score > 0. Since candidate translation 1 has no matching 4-gram, it has a BLEU score of 0.

The properties of the BLEU metric are as follows:

- **BLEU as a Corpus-based Metric**
  The BLEU metric does not performs accurately when used with single sentences. When bleu is computed for single sentences it captures less semantics rather than bleu computed for an entire corpus. The bleu metric defined cannot be used for individual sentences.
- **Less distinction between content and function words**
  The BLEU metric does not differentiate well between the stop words and the content words. If we drop a stop word 'the' and replace 'rover' with 'over' then the penalty would be same.
- **Not good at capturing meaning and grammaticality of a sentence**
  Dropping of negative words like not can change the meaning of the sentence. Bleu often impose a small penalty for grammatically wrong sentences.
- **Normalization and Tokenization**
  The tokenization and lemmatization a vital effect on the sentence quality. The sentences have to be normalized before computing the bleu value.

2.6 Calculating BLEU Score

1. The python NLTK library has BLEU metrics for comparing candidate sentence against reference sentence. NLTK provides sentence_bleu function for comparing the candidate sentence against reference sentences (Fig. 2.21).

The bleu score is 1 as the candidate sentence matches with the reference sentence.

NLTK provides corpus_bleu function for calculating bleu score for multiple sentences in the corpus. The reference documents are list of sentences having tokens.

```
1 from nltk.translate.bleu_score import sentence_bleu
2 reference = [['this', 'is', 'a', 'test'], ['this', 'is' 'test']]
3 candidate = ['this', 'is', 'a', 'test']
4 score = sentence_bleu(reference, candidate)
5 print(score)
```

Fig. 2.21 Bleu_score

2. An individual N-gram score is to match 1-gram or 2-gram (Fig. 2.22).
3. The cumulative n-gram scores are computed by taking geometric mean of the words in the sentence. Both the functions sentence_bleu and corpus_bleu are combined to compute the bleu score (Figs. 2.23 and 2.24).

```
1 # 1-gram individual BLEU
2 from nltk.translate.bleu_score import sentence_bleu
3 reference = [['this', 'is', 'small', 'test']]
4 candidate = ['this', 'is', 'a', 'test']
5 score = sentence_bleu(reference, candidate, weights=(1, 0, 0, 0))
6 print(score)
```

Running this example prints a score of 0.5.

```
1 0.75
```

We can repeat this example for individual n-grams from 1 to 4 as follows:

```
1 # n-gram individual BLEU
2 from nltk.translate.bleu_score import sentence_bleu
3 reference = [['this', 'is', 'a', 'test']]
4 candidate = ['this', 'is', 'a', 'test']
5 print('Individual 1-gram: %f' % sentence_bleu(reference, candidate, weights=(1, 0,
6 print('Individual 2-gram: %f' % sentence_bleu(reference, candidate, weights=(0, 1,
7 print('Individual 3-gram: %f' % sentence_bleu(reference, candidate, weights=(0, 0,
8 print('Individual 4-gram: %f' % sentence_bleu(reference, candidate, weights=(0, 0,
```

Running the example gives the following results.

```
1 Individual 1-gram: 1.000000
2 Individual 2-gram: 1.000000
3 Individual 3-gram: 1.000000
4 Individual 4-gram: 1.000000
```

Fig. 2.22 N-gram matching

```
1 # cumulative BLEU scores
2 from nltk.translate.bleu_score import sentence_bleu
3 reference = [['this', 'is', 'small', 'test']]
4 candidate = ['this', 'is', 'a', 'test']
5 print('Cumulative 1-gram: %f' % sentence_bleu(reference, candidate, weights=(1, 0,
6 print('Cumulative 2-gram: %f' % sentence_bleu(reference, candidate, weights=(0.5, 0
7 print('Cumulative 3-gram: %f' % sentence_bleu(reference, candidate, weights=(0.33,
8 print('Cumulative 4-gram: %f' % sentence_bleu(reference, candidate, weights=(0.25,
```

Fig. 2.23 Bleu score computation

```
Cumulative 1-gram: 0.750000
Cumulative 2-gram: 0.500000
Cumulative 3-gram: 0.000000
Cumulative 4-gram: 0.000000
```

Fig. 2.24 Cumulative n-gram score generation

2.7 Conclusion

Hence to evaluate the effectiveness of text classification methods different classifiers are tested on the available corpus and datasets.

1. We tried to implement a monolingual classifier for training and testing the corpus in the same language. The accuracy of this method is usually higher if the corpus is available in the single language.
2. We used Semi-supervised learning using model training and expectation maximization algorithm on 4000 labeled sentences and 1000 unlabeled sentences. The corpus here is error prone as it consists of un-annotated data as-well. The Expectation Maximization Algorithm outperforms the supervised learning approach on training data.
3. The Expectation Maximization model was applied on English-Chinese and English-French language pairs. It was seen that English-French sentence pairs are classified more accurately as compared to English-Chinese pairs (Fig. 2.25).

Category	English → Chinese														
	ML			MTS			MTT			EM			SEMI		
	P	R	F	P	R	F	P	R	F	P	R	F	P	R	F
sports	96.1	94.3	95.2	80.6	81.7	81.2	81.7	83.8	82.7	83.1	84.7	83.9	92.1	91.8	91.9
health	95.1	93.1	94.1	80.8	81.5	81.2	81.6	83.5	82.6	84.5	85.8	85.2	90.2	91.7	90.9
business	91.6	93.1	92.4	81.3	81.9	81.6	80.7	81.0	80.9	81.6	82.0	81.8	87.3	89.3	88.3
entertainment	88.1	88.3	88.2	76.1	78.8	77.5	75.3	78.9	77.1	76.8	79.7	78.2	83.2	83.8	83.5
education	79.1	82.2	80.6	70.2	72.5	71.8	71.1	72.0	71.6	71.2	73.7	72.5	76.2	79.8	78.0
	English → French														
sports	95.8	95.0	95.4	82.8	83.6	83.2	82.1	83.0	82.5	85.3	87.1	86.2	92.5	92.1	92.3
health	94.2	94.5	94.3	82.6	83.9	83.2	81.8	83.0	82.4	86.2	87.2	86.6	92.0	92.2	92.1
business	90.1	92.2	91.1	81.4	82.1	81.7	81.3	81.8	81.8	84.4	84.3	84.4	88.3	89.2	88.8
entertainment	87.4	87.2	87.3	76.6	79.1	77.8	76.0	78.8	77.4	78.9	81.0	80.0	84.3	85.5	84.9
education	78.8	81.8	80.3	72.1	74.8	73.5	72.3	72.7	72.5	73.8	76.2	75.0	76.3	80.1	78.2

Fig. 2.25 Comparison of classification methods

References

1. Strapparava, C., Mihalcea, R.: Semeval-2007 task 14_ Affective text. In Proceedings of the Fourth International Workshop on Semantic Evaluations (SemEval-2007), pp. 70–74 (2007 June)
2. Cendejas Castro, E.A.: Alineación automática de textos paralelos a nivel de palabras información lingüística diversa (Ph.D. thesis). Centro de Investigación en Computación-IPN, México (2013)
3. Amini, M.-R., Goutte, C.: A co-classification approach to learning from multilingual corpora. Mach. Learn. **79**(1–2), 105–121 (2010)
4. Guo, Y., Xiao, M.: Cross language text classification via subspace co-regularized multiview learning (2012). arXiv preprint arXiv:1206.6481
5. Wan, X.: Co-training for cross-lingual sentiment classification. In: Proceedings of the Joint Conference of the 47th Annual Meeting of the ACL and the 4th International Joint Conference on Natural Language Processing of the AFNLP: Volume 1-volume 1, pp. 235–243. Association for Computational Linguistics (2009)
6. Blitzer, J., McDonald, R., Pereira, F.: Domain adaptation with structural correspondence learning. In: Proceedings of the 2006 Conference on Empirical Methods in Natural Language Processing, pp. 120–128. Association for Computational Linguistics (2006)
7. Svenstrup, D., Hansen, J.M., Winther, O.: Hash embeddings for efficient word representations. In: Proceedings of the Advances in Neural Information Processing Systems 30 (NIPS 2017) (2017)
8. Johnson, M., Schuster, M., Le, Q.V., Krikun, M., Wu, Y., Chen, Z., Thorat, N., Viegas, F., Wattenberg, M., Corrado, G., et al.: Google's multilingual neural' machine translation system: enabling zero-shot translation (2016). arXiv preprint arXiv:1611.04558
9. Wick, M., Kanani, P., Pocock, A.: Minimally-constrained multilingual embeddings via artificial code-switching. In: Proceedings of the Thirtieth AAAI Conference on Artificial Intelligence, AAAI'16, pp. 2849–2855. AAAI Press (2016). http://dl.acm.org/citation.cfm?id=3016100.3016300
10. Mikolov, T., Chen, K., Corrado, G., Dean, J.: Efficient estimation of word representations in vector space (2013). arXiv preprint arXiv:1301.3781
11. Shi, L., Mihalcea, R., Tian, M.: Cross language text classification by model translation and semi-supervised learning. In: Proceedings of the 2010 Conference on Empirical Methods in Natural Language Processing, pp. 1057–1067. Association for Computational Linguistics (2010)

Chapter 3
Utilization of Internet of Things in Health Care Information System

Y. Harold Robinson, X. Arogya Presskila and T. Samraj Lawrence

Abstract The maintenance of Patient Information system is a huge challenge in the current scenario. The health care improvement demand the enhance efficiency for the service providers. Internet of things can utilize the services of the healthcare domain will increase the quality of the real life and maintains the helping hand for the professionals in this system for making the decisions. The information availability of the health care professionals can utilise the services of Internet of Things (IoT). The exchange of Health care related records can be easily available in the Internet of Things networks. The IoT is used to integrate all the smart devices, the available resources and the drugs prescription and to maintain the health related records in a single network. There are several challenges are facing by the healthcare organizations to exploit the services. This chapter addresses the problems which are facing in the healthcare sector. The interview methodologies with every kind of questions are examined and find the solution for this problem.

Keywords Internet of things · Health information · Exchange · Health information system · Utilization

---

X. Arogya Presskila
Department of Computer Science and Engineering, SCAD College of Engineering and Technology, Tirunelveli, India
e-mail: presskila@gmail.com

T. Samraj Lawrence
Department of Computer Science and Engineering, Francis Xavier Engineering College, Tirunelveli, India
e-mail: er.samraj@gmail.com

Y. Harold Robinson (✉)
School of Information Technology and Engineering, Vellore Institute of Technology, Vellore, India
e-mail: yhrobinphd@gmail.com

V. E. Balas et al. (eds.), *Internet of Things and Big Data Applications*, Intelligent Systems Reference Library 180, https://doi.org/10.1007/978-3-030-39119-5_3

3.1 Introduction

The Internet of Things (IoT) is the innovative technologies communicating the world through the dynamic and smart devices or objects for the reason of flawlessly gathering and sharing any type of data from anywhere, anytime and through any media from their surroundings [1]. The proposed utilization of IoT technology is to construct our lives elegant and secure with a exceptional identifier for every object [2]. In the healthcare system, IoT presents real-time contact to several types of health information rapidly and resourcefully, thus, providing convenience which is critical for contributing information within the healthcare officials. The IoT framework permits effortless access to and manage of the data [3]. In recent years, numerous devices have materialized for the purpose of exchanging data by using dissimilar technologies [4]. Most of these Technologies utilize the Internet as the bridging network to facilitate convenience of the information from the available resources [5]. In fact, the accessibility of health data has suit significant during the recent years [6]. Accessibility to information could be enormously helpful to decision makers, particularly when the accessible information is huge scale surveillance information [7]. Additionally, IoT might aid doctors in the judgment of health patient position and make possible suggested treatment with opportune interference. Furthermore, the connection within the healthcare contributes in healthcare deliverance is essential to attain attractive patient conclusion [8].

Within the network of IoT, there is the capability to implant a huge amount of devices to construct access to the large scale of data [9]. It can be viewed as a network of computers that can distribute software and information by using the internet [10]. The IoT network assists the assortment of entire data in the same network by using cloud computing to accumulate the data for easy monitoring of patient health updated details [11].

3.2 Related Works

Now-a-days, health information systems are being utilized by several healthcare providers, but there little technological and personality challenges among dissimilar hospitals. These obstructions comprise confrontation to data sharing and stoppage to recognize the latest techniques. Although these problems, the related work regarding the application of health related Information Technology and the internet for exchanging health data in together for developed countries reveals that the patient's satisfaction [12].

For the several years, countless countries initiated healthcare schemes designed at developing a countrywide interoperable Health Information System (HIS) but have come upon several problems such as the requirement of data exchange standards no established sustainable business representations and elevated risks in HIT speculations [13]. Previous researches established that the contribution of patient data within dissimilar healthcare providers is an unsurprisingly considerable challenge because of employee's resistance to the novel technique while the continuation of consistent framework for exchanging data is still lacking [14, 15]. As such a healthcare information system needs to ascertain a collaborative platform to split the required efforts in order to enlarge its receipt and use within dissimilar participants to increase healthcare competence and excellence.

The General Practitioners accounted that the need of health data led to sometimes needless, onerous tests and treatment. On the other hand, it looks the needs a longer period to recognize medical issues [16]. Additionally, healthcare provider's scheme is fragmented and the rough distribution of healthcare needs can be considered as the main problems [17]. Otherwise, deficiency of careers within the consultation and occurrence of part-time provisional care staff or insufficient recording and sharing within the colleagues may all guide to inadequate data that could be rescued for purposes of diagnosis and the formulation of management plans [18]. Due to conflicting interests among multiple stakeholders it is difficult to be reconciled and there is therefore, limited interoperability due to inconsistent standards, failure of financial incentives and an increase in healthcare demand for quality services, all of which pose a serious risk to under-diagnosis and under-treatment [19].

The data sharing with communication parameter is not working appropriately. This is owing to the constraints on successful management of care and this led to unsuitable healthcare and poses a risk to the unsurpassed possible care [20]. Hence, there is a requirement for logical and methodical interoperability within clinical data related systems to attain incorporation throughout flawless data transformation [21]. It is reported that construction of an e-health exchange system for medical information was an excellent suggestion to offer better incessant medical care and superior security to the health care system as well as the potential to control a number of laboratory tests and the number of doctor discussions [22].

The improvement in medical expenditure creates huge problems to the long-term process and feasibility of the Health management system. Comprehensive data for a patient detail is needed for data sharing and applying the health data exchange system is an inter-organizational procedure that difficulty the important integration of the personal and legal implications [23]. Hence, using such technique as IoT may assist the substitute of health data and construct it obtainable for any medical team that needs it. Additionally, several problems must find the solutions before adopting the HIE infrastructure in organizational and technological features which are vital the human feature like consumer acceptance [24]. The healthcare system is facing a lack of interoperability and association within the different models of hospital [25]. There is no exchange within most of these hospitals. There are only three large hospitals with the Health Information Exchange (HIE) system for persistent diseases that primarily shape the adult population [26].

3.3 Limitations for the IoT Based Healthcare Information System

A lot of challenges are needed to be addressed for implementing HIE which includes the urgent requirement for the patient regarding the medical sector. There will be the exchange of health data within the main and supplementary healthcare providers need to be organized with 100%. The workflow is sometimes interrupted for some specific issues, so this problem needs to be got the perfect solution. The user may collect every data using the latest technologies. Difficulty with finding qualified and trained staff and a number of part-time clinicians augment coordination problems that produces the modest utilization based workflow model and unused technologies of HIE. Working with the individual vendor organizations must be needed for superior teamwork and incorporation efforts, exposure technology error and training. Moreover, the apparent communication and data sharing with General Practitioners (GPs) not to work healthy in the care procedure. It is very hard to acquire or occupied historical data relevant to patient data including medical histories and testing results with enhanced amount of disaster visitors. Moreover, different providers have varying data requirements, users access HIE systems in different ways within varies organizations. The HIE its not extensive adopted and incomplete in convenience for health patient data. The adoption of an HIE needs to full understanding of the latest technology benefits. The patient details exchange within the hospitals still has scientific and character barriers including resistance to information sharing. The major challenge is employee's resistance and also the capabilities of data storage for doctor consolations sympathetic benefit and user objective restricted to adoption and many others factors are effects of using of health data exchange.

Regional HIE in premature period due to the requirement of interoperability standards and accomplishment strategy. The healthcare authorities working on whole the building of health exchange within the quick period of time but still need to increase the competence and sustainability of finished systems. The healthcare has a facing require in the interoperability and the relationship within all kind of hospitals. There is no exchange within the hospitals. Conversely, only minimum amount of hospitals have Health Information Exchange (HIE) system for chronic diseases that facing the adult population. Another problem is the short of the studies and government roles and also the minimum amount of agreement on what EHRs capabilities mean and comprise. T0 execute and accept of EHR does not achieve the preferred rate of allotment. The Poor hospital management issues and user confrontation will be using the latest technology and no revival planning. Lacks in using HIE and no any significant correlations between the HIS. The requirement of the Patient and the agreement to solve the problems has some amount of limitations.

3.4 Materials and Methods

This methodology is implemented in order to acquire the information on utilization of IoT services to foster the health information system. This methodology aimed to establish the primarily factors related to the utilization of IoT services for health information exchange. The researcher used the interview methodology with a semi structured mechanism to accomplish in-depth understanding of the subject in question. The IoT-based healthcare system offers multiple advantages as summarized in the following points.

Remote patient monitoring continues in order to help physicians in their diagnoses and treatment of illnesses and diseases by obtaining dependable data within a negligible error rate. IoT with intelligent medical sensors increases the quality of life considerably and averts the occurrence of health problems. Minimize the unnecessary visits to the doctor and readmissions for those patients who have chronic diseases and thus reduce testing and treatment costs. Flexibility and mobility is helped in accessing data anywhere, anytime and in any media. Increases the care quality and control by enhancing the management of drugs, reduces the medical error, enhances the patient experience, improves the disease management and improves treatment outcome. As illustrated in Fig. 3.1 that the Cisco Revolution defined it as the internet of everything: people, processes, data and things.

The system Enables automatic information gathering from health resources such as monitoring, first aid, tracking, analysis, diagnosis, alarm-triggering, locating and collaboration with medical healthcare on a unified communication platform. It then facilitates an interaction among the parts of an enterprise that reduces the time required to adapt itself to the changes demanded by the market evolution.

Massive health data in IoT, devices assemble and communicate data directly together through the internet as well as cloud which manages in order to gather records and analyze the required information in the system. However, tune things or devices that construct huge amounts of data flow on a day-to-day basis and needs to be treated and managed. Even with the above drawbacks, the IoT still offers many benefits to the health care system. It provides data in real time of health position of patients and also data to doctors who assist. Moreover, IoT facilitates the utilization of communication forms which will grow from human-human to human-thing and thing-thing. That means, IoT forever offers innovative tools which will encourage efficiencies to realize the integration of healthcare systems in the circumstance of modem healthcare to guarantee appropriate care delivery to patients. This will diminish the healthcare expenditure and expand treatment outcomes. In light of the above, there is a requirement to approve IoT technology to make possible the HIE among the

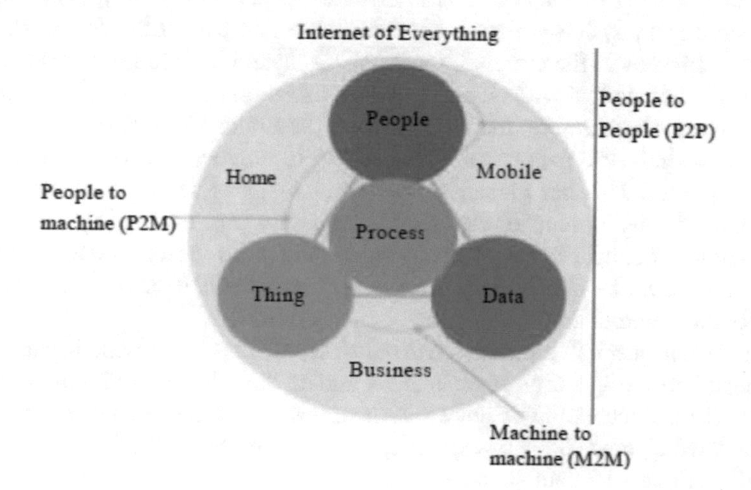

Fig. 3.1 Internet of things revolution

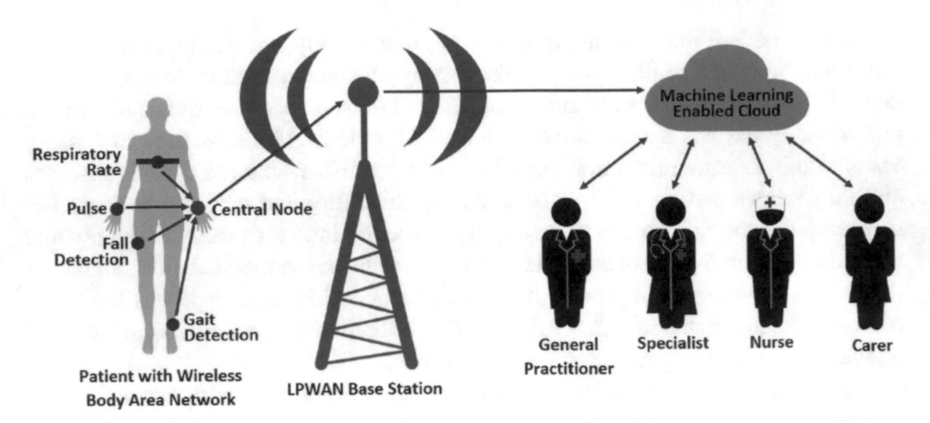

Fig. 3.2 Health information system

hospitals and improve the healthcare quality by being able to monitor health patient position. Figure 3.2 demonstrates the Health Information system with the concept of Internet of Things. The Base station is connected with the patient information system and the Machine learning enabled cloud.

When considering this, efforts have been completed by the Health authorities to exploit specific ubiquitous technology models to increase health-related medical practices in different sectors. This involves re-engineering the performance of access to health information which would permit healthcare professionals to observe and understand patient's conditions efficiently. Numerous organizations have failed to exploit innovative technology such as internet of things. The need of understanding of the needs of healthcare organizations to exploit innovative technology can obstruct the utilization of such technology in the healthcare system. Therefore, efficient exploitation of IoT technology in hospitals would have to address the resources needs so as to create the highest possible assessment and evade failure. Currently the system in most of the hospitals is not able to access patient medical records in real-time and physicians have difficulties accessing patient health records from anywhere. Moreover, the current circumstances make it difficult to adopt a certain technology such as IoT without any prior examination of its suitability within the usage context. Furthermore, the HIE includes dissimilar kinds of procedure based on different kinds of organizations that prolong to improve the access stipulate for health resources. This has given rise to various major issues for the public health care sectors during 3 points of view; low IT integrity, data exchange procedure and information difficulty. Figure 3.3 demonstrates the pulse sensor which is located in the human body. The radial artery is connected with the LED and Photodiode with some Radius value.

The selected hospitals are gathered in built-in the focus of the research, particularly with regard to the quantity of physicians and IT practitioners who are the most frequent users of technology. Figure 3.4 illustrates the Pressure related sensor which is located in the human body by extracting the force from the pulse with radial artery and radius. Figure 3.5 demonstrates the PPG Sensor.

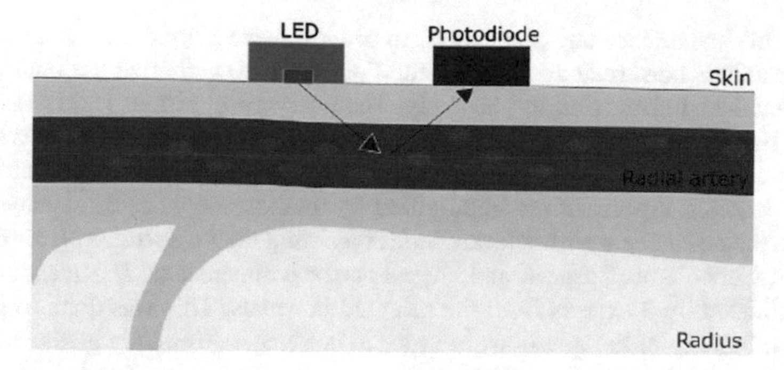

Fig. 3.3 Pulse sensor

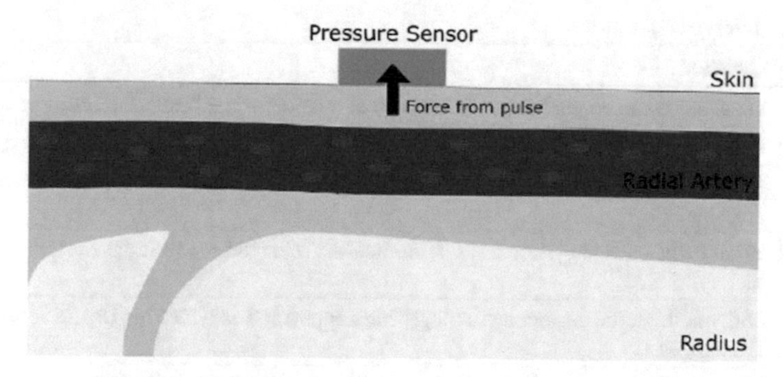

Fig. 3.4 Pressure related sensor

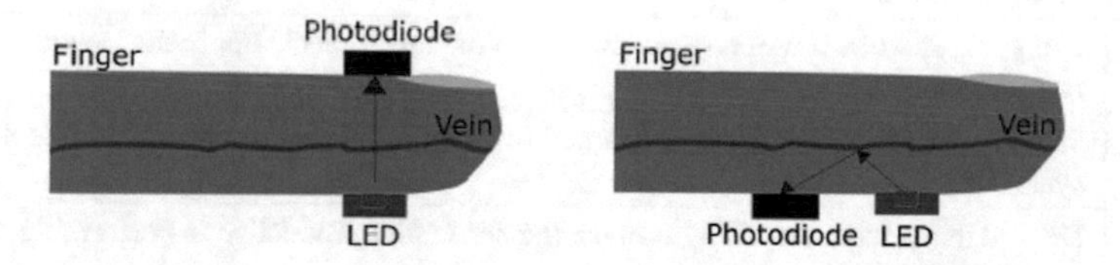

Fig. 3.5 PPG sensor

The other hospitals were excluded because they did not meet the criteria with respect to resources and time. The physicians and IT practitioners of the chosen hospitals had knowledge of IoT services. Of the 29 interviewees, 14 were IT practitioners and 15 of them Physicians. The data collection method is used was face-to-face interview and the researcher used the semi-structured form. The demographic profiles of the interviewees were as follows; were aged ranged between 26 and 49 years; seven were female and 22 were male years of experience ranged from 1.5 to 18 years in using current technology in health information systems.

The interview is a common and popular instrument used for data compilation in qualitative research. The qualitative method was used to focus in-depth on the individual's experience and perception. In order to have a direct for the interviews, the researchers constructed 15 open-ended questions to authorize the interviewees to current deep information for the study. The interview questions were constructed guided by the concerns of the Ministry of Health about the use of IoT services for HIE among hospitals. The interview questions accumulated also took into deliberation assorted aspects of the highlighted by the hospital's health managers who assisted the researchers with relevant input regarding their respective organizational systems, technological, aspects and several personal dimensions. The questions were then validated by 3 experts from the selected hospitals. This was done to guarantee the reliability of the questions in order to congregate the study goals. Table 3.1 demonstrates the set of 15 questions.

Table 3.1 Interview questions

S. No.	Question
1	What are the main challenges facing the utilization of the health information system?
2	Did you think IoT services can improve the Health Information Exchange system?
3	Do you see the exchange of health information through the current system as easy and possible?
4	What is the current system used in the health sector to keep past patient treatment history?
5	Did you think the current technology can support IoT services for Health Information Exchange?
6	What are the main challenges facing staff involved in the information exchange?
7	Do you have access to online technical support and is you do is it available to you in a timely manner?
8	Did you think that it will be easy for health organizations to apply the Health Information Exchange system?
9	What do you think should be developed to support the Health Information Exchange system?
10	Do you think using the IoT will make the staff more productive in performing their work?
11	What is the current technology used in Health Information Exchange between organizations?
12	Does the existing Health Information Exchange management system work as well as you expected?
13	Personally, what do you think are the antecedents that could impact the use of IoT services?
14	Do you have a patient-central for your hospitals?
15	What is the main challenge you face in using IoT services in your hospital?

3.5 Results and Discussion

Decades ago, the health status had suffered from progressive neglect and low budgetary allocations because of other issues. The Ministry of Health is now under pressure to reinstate lost impetus and expand the healthcare systems. The researcher conducted interviews to emphasize the present needs in hospitals and to examine the most important antecedents that face the consumption of IoT in these hospitals. The interview questions were assembled based on the problems raised by the health information system's decision makers to exploit HIE among the hospitals.

During the interviews, the interviewees provided in-depth answers regarding the exploitation of Health Information System and the status of information exchange within dissimilar healthcare providers. The answers revealed by the 29 interviewees painted a negative picture of the utilization of present technologies in the Health Information Systems. Some of the hospitals were still using the paper-based system although having electronic health systems; this was due to the user's cynicism concerning the modern technologies used and the security of the information. Additionally, the interviewees reported some problems in using the health information systems such as the lack of health patient information sharing, difficulty in acquire previous treatment records for patients and the low ratio of number of physicians to patient's admissions and the patient's understanding.

In the Emergency Department, the interviewees reported problems in accessing the health position of the patient on admission and they had to accomplish new tests and construct latest diagnoses. Also, the increased amount of received patients in the emergency department has unavoidably improved the time taken for admission which has exasperated patients and led to public disappointment with the service at the Emergency Department. The interviews with 2 heads of Emergency Departments from different hospitals revealed that the absence of an HIE system was a challenge in terms of controlling and monitoring the health workers.

The interviewees, also, complained about the difficulty of accessing the information. Moreover, they admitted it would be much easier to work with the HIE system. Most of the respondents agreed that the existing system did not work as planned. Therefore, the respondents indicated that they like to see the system being upgraded to fit their workflow. A different system is used to develop the clinical data which makes it difficult to process or update the information. The respondents mentioned that there were no flowcharts available to keep track of data. Therefore, the understanding of the user's workflow is based on process requires and as such it is needed to implement a system that accommodates these needs.

With a view to utilizing a new system, the interviewees expressed concern about the cost effectiveness because of the deficiency of a business plan. These concerns were due to the restricted budget of the organization to assemble all the system requirements and also based on fears about enhance in the expenditure of maintenance of hardware and software availability. Other researchers have indicated that cost saving is a benefit of using the IoT-based healthcare system.

The interviewees in the IT department designated that they would like to approve latest technology with most of the health devices having the capability to be communicated through the internet. It was also, reported that most of the medical staff had poor computer knowledge and therefore, their record-keeping was primarily paper-based. It was also regular practice to use CDs, USB memory sticks and email for data exchange within the hospitals. As such sharing of patient data with other hospitals was neither suitable nor possible. The researchers, also, reported poor internet connectivity is compared too many other countries. The participants attributed the poor internet condition to the cost factor. On the complete it was normally agreed that IoT technology would considerably increase the present healthcare data exchange systems effectiveness if all these related issues could be addressed.

As a result were found during the survey of the interview. Controlling the progress in using IoT technology in health sectors were found to be due to some reasons related to numerous factors. The participants complained from increasing the workflow when using both electronic and paper-based data entry. In addition, they also highlighted other concerns related to the information security and privacy when exploiting the IoT technology.

After a number of factors realize, the researcher administered the agreement form to the interviewees to asking them about their agreements upon for the extracted factors that they mostly agreed upon. The responses were combined in Table 3.2. From the result it might be concluded which majority of the respondents agreed on these factors and that they preserve a remarkable impact on current utilization of IoT services for health information exchange.

Table 3.2 Agreements on the extracted factors

Factors	Agreements
Organizational domain	
Workflow	26
Cost-effectiveness	29
Cooperation	26
Training	29
Technological domain	
Security and privacy	28
Ubiquitous connectivity	27
Compatibility	28
Network capacity	29
System domain	
Accessibility	29
Usefulness	26
Individual domain	
Actual usage behaviour	25
Trust	25

3.6 Conclusion

Additionally there are also, factors such as accessibility and usefulness that need to be considered as they would affect personal factors associated with the actual usage behaviour and trust in the system. From the results it can be concluded that the majority of the interviewees agreed these identified factors greatly impacted the current utilization of IoT services in the health information systems.

References

1. Ahrnadian, L., Nejad, S.S., Khajouei, R.: Evaluation methods used on health information systems (HISs) in Iran and the effects of HISs on Iranian healthcare: a systematic review. Int. J. Med. Inform. **84**, 444–453 (2015)
2. Campion, T.R., Edwards, A.M., Johnson, S.B., Kaushal, R.: Health information exchange system usage patterns in three countries: Practice sites, users, patients and data. Int. J. Med. Inform. **82**, 810–820 (2013)
3. Harold Robinson, Y., Balaji, S., Golden Julie, E.: FPSOEE: Fuzzy-enabled particle swarm optimization-based energy-efficient algorithm in mobile ad-hoc networks. J. Intell. Fuzzy Syst. **36**(4), 3541–3553 (2019)
4. Dobrzykowski, D.D., Tarafdar, M.: Understanding information exchange in healthcare operations: evidence from hospitals and patients. J. Oper. Manage. **36**, 201–214 (2015)
5. Downing, N.L., Adler-Milstein, J., Palma, J.P., Lane, S., Eisenberg, M., et al.: Health information exchange policies of 11 diverse health systems and the associated impact on volume of exchange. J. Am. Med. Inform. Assoc. **24**, 113–122 (2016)
6. Balaji, S., Golden Julie, E., Harold Robinson, Y., Kumar, R., Thong, P.H., Son, L.H.: Design of a security-aware routing scheme in mobile ad-hoc network using repeated game model. Comput. Stand. Inter. **66**, (2019)
7. Hameed, R.T., Mohamad, O.A., Hamid, O.T., Tapus, N.: Design of E-Healthcare management system based on cloud and service oriented architecture. In: Proceedings of the 2015 Conference on E-Health and Bioengineering (EHB'15), pp. 1–4, 19–21 Nov 2015. IEEE, Iasi, Romania (2015). ISBN: 978-1-4673-7544-3
8. Hekrnat, S.N., Dehnavich, R., Behmard, T., Khajehkazemi, R., Mehrolhassani, M.H., et al.: Evaluation of hospital information systems in Iran: a case study in the Kerman Province. Glob. J. Health Sci. **8**, 95 (2016)
9. Kadhurn, A.M., Hasan, M.K.: Assessing the determinants of cloud computing services for utilizing health information systems: a case study. Int. J. Adv. Sci. Eng. Inform. Technol. **7**, 503–510 (2017)
10. Latif, A.I., Othman, M., Sulirnan, A., Daher, A.M.: Current status, challenges and needs for pilgrim health record management sharing network, the case of Malaysia. Int. Arch. Med. **9**, 1–10 (2016)
11. Balaji, S., Golden Julie, E., Harold Robinson, Y.: Development of fuzzy based energy efficient cluster routing protocol to increase the lifetime of wireless sensor networks. Mobile Netw. Appl. **24**(2), 394–406 (2019)
12. Santhana Krishnan, R., Golden Julie, E., Harold Robinson, Y., Kumar, R., Son, L.H., Tuan, T.A., Long, H.V.: Modified zone based intrusion detection system for security enhancement in mobile ad-hoc networks. Wirel. Netw., 1–15 (2019)
13. Mastebroek, M., Naaldenberg, J., Mareeuw, F.A., Leusink, G.L., Lagro-Janssen, A.L., et al.: Health information exchange for patients with intellectual disabilities: a general practice perspective. Br. J. Gen. Pract. **66**, e720–e728 (2016)

14. Harold Robinson, Y., Rajaram, M.: Energy-aware multipath routing scheme based on particle swarm optimization in mobile ad hoc networks. Sci. World J., 1–9 (2015)
15. Harold Robinson, Y., Rajaram, M.: A memory aided broadcast mechanism with fuzzy classification on a device-to-device mobile ad hoc network. **90**(2), 769–791 (2016)
16. Melby, L., Brattheirn, B.J., Helleso, R.: Patients in transition-improving hospital-home care collaboration through electronic messaging: providers' perspectives. J. Clin. Nurs. **24**, 3389–3399 (2015)
17. Mishuris, R.G., Yoder, J., Wilson, D., Mann, D.: Integrating data from an online diabetes prevention program into an electronic health record and clinical workflow, a design phase usability study. BMC Med. Inform. Decis. Mak. **16**, 1–13 (2016)
18. Harold Robinson, Y., Golden Julie, E., Saravanan, K., Kumar, R., Son, L.H.: DRP: Dynamic routing protocol in wireless sensor networks. Wirel. Pers. Commun., 1–17, Springer (2019)
19. Richardson, J.E., Ves, J.R., Green, C.M., Kem, L.M., Kaushal, R., et al.: A needs assessment of health information technology for improving care coordination in three leading patient-centered medical homes. J. Am. Med. Inform. Assoc. **22**, 815–820 (2015)
20. Rinner, C., Sauter, S.K., Endel, G., Heinze, G., Thwner, S., et al.: Improving the informational continuity of care in diabetes mellitus treatment with a nationwide Shared EHR system: estimates from Austrian claims data. Int. J. Med. Inform. **92**, 44–53 (2016)
21. Harold Robinson, Y., Santhana Krishnan, R., Golden Julie, E., Kumar, R., Son, L.H., Thong, P.H.: Neighbor knowledge-based rebroadcast algorithm for minimizing the routing overhead in mobile ad-hoc networks. Ad Hoc Netw. **93**, 1–13 (2019)
22. Thannalingam, S., Hagens, S., Zelmer, J.: The value of connected health information: perceptions of electronic health record users in Canada. BMC Med. Inf. Decis. Mak. **16**, 93–100 (2016)
23. Harold Robinson, Y., Golden Julie, E.: MTPKM: Multipart trust based public key management technique to reduce security vulnerability in mobile ad-hoc networks. Wirel. Pers. Commun. **109**, 739–760 (2019)
24. Wang, J.Y., Ho, H.Y., Chen, J.D., Chai, S., Tai, C.J., et al.: Attitudes toward inter-hospital electronic patient record exchange: discrepancies among physicians, medical record staff and patients. BMC Health Serv. Res. **15**, 264 (2015)
25. Wu, H., Larue, E.: Barriers and facilitators of health information exchange (HIE) adoption in the United States. In: Proceedings of the 48th Hawaii International Conference on System Sciences (HICSS'15), pp. 2942–2949, 5–8 Jan 2015. IEEE, Kauai, Hawaii (2015). ISBN: 978-1-4799-7367-5
26. Zhang, H., Han, B.T., Tang, Z.: Constructing a nationwide interoperable health information system in China: the case study of Sichuan Province. Health Policy Technol. **6**, 142–151 (2017)

Chapter 4
IoT, Big Data, Blockchain and Machine Learning Besides Its Transmutation with Modern Technological Applications

G. Arun Sampaul Thomas and Y. Harold Robinson

Abstract Amid the augmentation of the Internet of Things (IoT), applications have become smarter and coupled devices give escalation to their exploitation in all facets of a modern city. As the capacity of the collected data increases, Machine Learning (ML) methods are applied to auxiliary boosting of intelligence and the abilities of an application. The field of smart transportation has fascinated many researchers and it has been accosted with both ML and IoT techniques. In this evaluation, smart transportation is contemplated to be a canopy term that conceals the route optimization, accident prevention/detection, parking, street lights, road anomalies, and infrastructure applications. The purpose of this document is to make a self-contained evaluation of ML techniques and IoT applications in Intelligent Transportation Systems (ITS) and attain a clear view of the developments in the above-mentioned fields and spot possible coverage musts. From the reviewed articles it becomes insightful that there is a possible lack of ML coverage for the Smart Lighting Systems and Smart Parking applications. Additionally, route optimization, parking, and accident/detection tend to be the most popular ITS applications among researchers, henceforth there is a huge possibility in implementing the IoT with real world applications with the support of various Big data concepts and Machine learning algorithms along with Block chain technology.

Keywords Internet of Things · Artificial intelligence · Machine learning · Smart transportation system · Intelligent transportation systems · Big data · Block chain

G. Arun Sampaul Thomas (✉)
CSE Department, J.B. Institute of Engineering and Technology, Hyderabad, India
e-mail: arunthomas.cse@jbiet.edu.in

Y. Harold Robinson
School of Information Technology and Engineering, Vellore Institute of Technology, Vellore, India
e-mail: yhrobinphd@gmail.com

V. E. Balas et al. (eds.), *Internet of Things and Big Data Applications*, Intelligent Systems Reference Library 180, https://doi.org/10.1007/978-3-030-39119-5_4

4.1 Introduction

The Internet of Things (IoT), big data and machine learning are three of the most exciting new business technologies of the past 5 years [1]. They have been the stalwarts of Gartner's hype cycle and yet many people don't really know the difference between them, often overlapping their uses and confusing their purposes.

With these three processes and systems, there are huge overlaps and they are also more or less symbiotic with one another. The future of technology will actually be driven by these three elements in their own ways.

What Is The IoT?

The IoT is a link made between connected devices. It allows devices to work outside of the need for human input.

To put it into a context that we can all understand, the IoT allows your fridge to interact with a pot of yoghurt inside it. Without any kind of intervention from a human, the fridge will be able to see when the expiration date of the yoghurt is, then once it is past that date, it will automatically inform the human and order a new pot.

It goes well beyond this though, allowing pipelines to feedback their status, cars to automatically book themselves in for a service on a particular part or even for thermostats to control temperature to optimize energy savings.

What Is Big Data?

Big data in itself is nothing, that is to say that it is more of a process than a thing.

It has become the umbrella term for the collection, analysis and storage of vast amounts of data. With the advent of the internet and a society that naturally creates a significant amount of data, it has meant that organizations can collect data on almost anything in huge quantities. Big data is the act of collecting and storing this data.

This can then be used for other things, like analytics, where the data collected is run through specific algorithms in order to find certain insights or outcomes that may not be obvious from simple analysis or investigation.

What Is Machine Learning?

Machine learning is the ability of technology to create algorithms themselves to find a particular pattern or conclusion and then act upon it. This itself can then be analysed and acted upon, creating a cycle of learning and acting to create accurate and efficient processes.

So How Are They All Connected?

At present the IoT is small in comparison how big it will eventually become, big data is a fairly well entrenched new business principal and machine learning is in a similar position to the IoT. Each are going to grow in popularity further than they already have, but they will also need the other two to grow alongside them in order to create a symbiotic environment.

Devices connected to the Internet of Things will create huge amounts of data which will all be collected and stored. This will be put into useable formats and silos by big data. Machine learning will then use these huge oceans of data to improve processes and increase self-sufficiency of systems. These processes are then fed back into the devices connected to the IoT and the process can start again.

Each part has its own unique qualities that will be put to use outside of this particular cycle, the IoT in closed systems, big data in almost anything and machine learning in the automation and management of huge systems. However, in order for this particular system to work to its full potential, there needs to be co-operation and large amounts of growth in each of the segments.

So Are They The Same Thing?

Yes and no. In certain situations they create a process around themselves, each requiring the other in order to function, but at the same time can work separately for other purposes. It is often why the concepts are confused, they are all part of the same ecosystem whilst other business, social or organizational needs can be addressed with one or all of the parts.

4.2 Necessity of Only AI or IoT or ML or Blockchain or All of Them Together?

Few thought-provoking articles I had read recently about Machine-Joking. Humans had some achievement in defining what would create humour [2]. Can machines do that by means of AI with the help of ML? Nah, it's not been entirely codified. As even full-time improvised comedians would confess, there is no magic formulation to fabricate the unspoiled joke. Much of what creates us laugh depends on elusive elements such as context or gesture of the comedian himself. Sometimes even we humans don't know why a joke is comical!!! What is comical for one person may not be entertaining for others. The same joke told by two people might generate different responses. A joke is fun, and a jest can be subjective. So, how can we teach AI to make jokes if we ourselves don't grip the reasons why a joke is humorous? AI, which inclines to accentuate on a very slender range of tasks, is poorly furnished to spot the wide range of elements involved in humour, let alone know what they mean. Drought of context is one area that makes it more challenging for algorithms or codification.

Refer the following examples of the machine generated one-liners jokes:

- I enjoy my coffee like I enjoy my war: cold.
- You know what really thrusts my buttons? That guy that's in control of me.
- And like my Alexa at home told me a joke "What is the preferred crisp flavour of the pilot, answer is "Plane".

That in principle describes deficiencies of machine controlled, code-based undertakings. The system learns to deliberate from data. It is based on the trends and patterns from the data that the AI makes assessments. The assessments have to be logical and defined. AI is just basically an efficient assessment based decision-making machine. It is can be taught to think but it cannot generate thought beyond the patterns and trends from the data it receives.

Blockchain solutions:

- Blockchain is redefining how trustworthy transactions ought to be conceded out. The internet is itself highly susceptible and Blockchain is out with the result to discourse it. One problem that Blockchain resolves of AI and IoT is the confidence liability lines. Most IoT devices are associated to each other via public networks and it is inessential to say how susceptible public networks really are. Blockchain resolves this problem by linear and enduring indexed records which can be crafted [10]. Globally general public can mention them without censorship. They can also smoothen the trade process by providing a payment mechanism as well as communication outlet. The public is the authority and not any centralized entity as is the case with the banks. Any kind of hacking and tampering with the data like captivating control of device and records is intolerable due to the way blocks are stored and guarded in an unambiguous database in the Blockchain scheme. Every IoT device is a point of vulnerability and the hazards are still higher as even AI is involved in making choices for users. Hence Blockchain can be used to provide a safe, scalable and certifiable platform that has supreme security implementations.

Block Chain and its Process flow:

- Internet of things (**IoT**) (Sensors to evidence various task statistics, with **Blockchain** elucidations) **==> BigData** (Capability to store large volume of data, whether from sensors or from systems) **==>** Machine Learning (**ML**)/Deep Learning (**DL**) (Decision Options for AI based on design of data and statistics, originated from BigData) **==>** Artificial Intelligence (**AI**) (The decision maker, who decided grounded on best-case scenarios) as indicated in the Fig. 4.1.

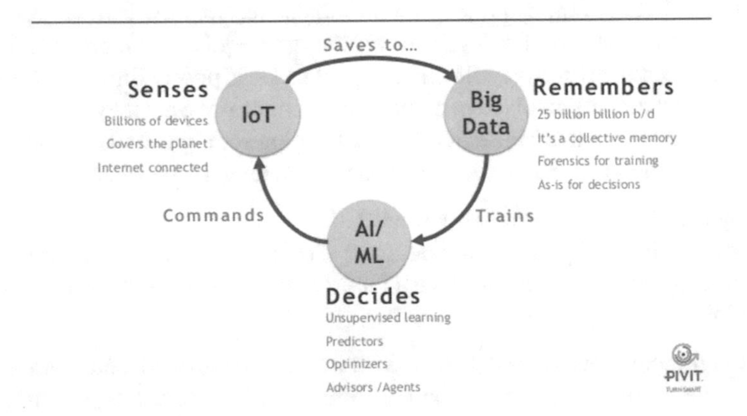

Fig. 4.1 Block chain solutions with process flow

4.3 Various AI Systems and Its Relationships

The tenacity of creating the AI systems is for optimization and proficiency. If they are predisposed to the same issues as humans, then they are not needed. An AI that can make a joke is like an AI that can make errors; not just conscious errors or errors but unconscious faults that are not errors. That downfalls the purpose of AI. Parenthetically it's one of the things that creates human's standout i.e. we do errors and we learn from our errors. And certainly, that is where art comes from. AI was built to be perfect, to have one that can-do jokes is to purposely build a defective one. But can it automatically learn from the errors like we humans do? How long that learning process takes for it to be flawless, where it does not make any more errors? Are these learnings for one machine can benefit other machine or every machine has to go through its own learning cycle? Or there will be a huge global data lake (Data Ocean may be); where all machine can take their intelligent feed from, to be at similar level as other machines? How long that maintenance process takes, for machines to impress over humans? That's the fear we are living today, right?

AI has its abode, but there are just some lines it cannot cross. This is not just because the system cannot learn it, rather it is because the system has to be anti-intelligent to attain it. Anti-intelligence will take AI away from its standout point. And if an AI can make a optimal solution whether to be intelligent or stupid, then the world is in misfortune; what humans have not achieved in thousands of years will be attained in days! It is possible, but better not explored. What we need for that is for AI to have a really good prototype of the world, which it doesn't have at this instant. Maybe this knowledge would contain knowing about feelings, human feelings, animal feelings, instigate and effect in all or most sense conceivable.

Today Artificial Intelligence, Machine Learning and Deep Learning are at the heart of digital revolution by empowering organizations to influence their growing prosperity of big data to enhance key business outcomes and spread operational used cases.

When the Internet became predictable, a lot of business models were flustered. Companies had to transmute themselves or vanish and we were privy to many examples such as bookstores, retailers, DVD sellers etc. Rather similar is about to happen with AI in the next few decades. Nevertheless, with a big dissimilarity: AI is not going to be a new industry; it is going to be in every industry. It's going to be in every application, in every process imaginable, and in every aspect of our lives, almost unsparingly. It won't just remain gratified with the business world and will increasingly encroach on further areas like culture and art, just as much. It's our responsibility to make certain that this new revolution turns out right, our number one priority should be to make intelligent and cognitive capabilities available to everyone. Considerable like the ease of creating a web app today because the tools and technologies underlying the web are easy to use, have open-source, and are usually free. In addition, learning resources are readily available for free or at a throwaway price. Of course, this is not about transitioning everyone to jobs that will involve AI, rather this is about making sure that all those who have the potential to create value with AI are able to do so freely. This is about guaranteeing that no human potential goes to waste. And that is possible; first by creating awareness, encouraging discussions, creating knowledge, mass-circulate the knowledge, creating alternative occupations or opportunities for those who are affected and then finally landing the prospect—a sustained and step by step method, with the collaboration of government and private sectors.

First thing first; let's start with some of the most commonly used acronyms and their definitions:

- **Artificial Intelligence (AI)**—is the overarching discipline that covers anything related to making machines clever. Whether it's a robot, a refrigerator, a car, or a software application, if we are making them clever with smart thinking, then it's AI.
- **Machine Learning (ML)**—is commonly used alongside AI but they are not the similar thing. ML is a subset of AI. ML refers to systems that can learn by themselves. Systems that get smarter and cleverer over time without human intervention.
- **Deep Learning (DL)**—It is machine learning method but applied to large data sets.
- **Artificial Neural Networks (ANN)**—Refers to prototypes of human neural networks that are designed to help computers learn. *There are many techniques and methods to Machine Learning. One of those approaches is artificial neural networks (ANN), occasionally just called neural networks*. A good example of this is Amazon's recommendation engine. Amazon uses artificial neural networks to produce recommendations for its customers. Amazon suggests products by showing us "customers who viewed this item also viewed" and "customers who bought this item also bought". Amazon Inc. espouses data from all its users browsing experiences and uses that information to make effective product recommendations.
- **Natural Language Processing (NLP)**—Refers to systems that can recognise language. NLP is the processing of the text to recognise meaning.
- **Automated Speech Recognition (ASR)**—Refers to the use of computer hardware and software-based techniques to identify and process human voice i.e. managing of speech to text. Because humans speak with idioms and abbreviations it takes wide computer analysis of natural language to initiate accurate outputs.

As per Fig. 4.2**, following features are framed**,

- ASR and NLP plunge under AI.

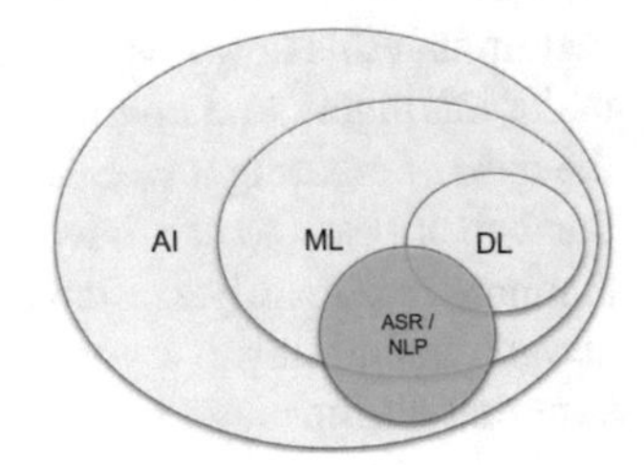

Fig. 4.2 AI versus ML versus DL with NLP processing

- ML and NLP have some intersection as ML is often used for NLP tasks.
- ASR also intersections with ML. It has historically been a driving force behind many machine learning techniques.
- Most AI work now contains ML because intelligent behaviour requires considerable knowledge, and learning is the easiest way to get that knowledge.
- Data Analytics, Predictive Analytics and Prescriptive Analytics are different applications of Artificial intelligence.

That is the relationship between AI & ML. The image below captures the relationship between AI, ML, and DL.

AI encompasses machines that can perform tasks that are characteristic of human intelligence. While this is rather general, it comprises things like planning, understanding language, recognizing objects and sounds, learning, and problem solving.

We can position AI in two categories, general and narrow. General AI would have all of the features of human intelligence, including the dimensions mentioned above. Narrow AI exhibits some facet(s) of human intelligence, and can do that facet enormously well, but is lacking in other areas. A machine that's great at recognizing images, but nothing else, would be an example of narrow AI.

At its core, machine learning (ML) is merely a way of attaining AI. We can get AI deprived of using machine learning, but this would oblige building millions of lines of codes with complex rules and decision-trees. So in its place of hard coding software routines with specific instructions to accomplish a particular task, machine learning is a way of "training" an algorithm so that it can learn how. "Training" involves feeding enormous amounts of data to the algorithm and allowing the algorithm to adjust itself and improve. It reminds me for the Man and Fish story: "*Give a Man a Fish, and we Feed Him for a Day. Teach a Man to Fish, and we Feed Him for a Lifetime*". To give an example, machine learning has been expended to make drastic improvements to computer vision (the ability of a machine to recognize an object in an image or video). We congregate hundreds of thousands or even millions of pictures and then have humans tag them. For example, the humans might tag pictures that have a cat in them versus those that do not. Then, the algorithm tries to build a model that can accurately tag a picture as encompassing a cat or not as well as a human. Once the accuracy level is high enough, the machine has now "learned" what a cat looks like. I would imagine, a tag proposal from Facebook is built on same or similar principle.

Deep learning is one of many tactics to machine learning. Other approaches embrace decision tree learning, inductive logic programming, clustering, reinforcement learning, and Bayesian networks, among others. Deep learning was stimulated by the structure and function of the brain, namely the intersecting of many neurons. Artificial Neural Networks (ANNs) are algorithms that mimic the biological structure of the brain. In ANNs, there are "neurons" which have discrete layers and connections to other "neurons". Each layer picks out a precise feature to learn, such as curves/edges in image recognition. It's this layering that gives deep learning its name, depth is generated by using multiple layers as opposed to a single layer.

4.4 How to Apply Machine Learning Algorithms to IoT Data?

The entire earth is going crazy over data, IoT and Artificial Intelligence. Lots of articles have spoken about the amount of data we generate every single day and numerous statistics have shown how much data we would generate by the year 2025. On this post, however, we are going to deviate a little from data generation and deliberate how algorithms or concepts from other technologies would be applied to IoT data for optimizations [4]. On one of our previous posts, we discussed Data Science algorithms with IoT data and today, it will be Machine Learning.

Machine Learning became a household term when Facebook shut down it's Artificial Intelligence division when one of its bots discovered a whole new language. With Elon Musk commenting on it and netizens indicating an I-Robot in the future, most of us understood what Machine Learning all is about.

On a very basic intellect, machine learning in technology today is the process of elimination of human interference wherever possible. It is allowing the data to learn patterns by itself and take autonomous decisions without a coder having to write a new set of codes. If you use your Siri, for instance, you would notice that its responses are more refined and appropriate as you keep using it. That is one of the simple applications of Machine Learning.

But when it comes to a complicated concept like IoT, how would Machine Learning make things better for the Internet of Things? Every time the IoT sensors gather data, there has to be someone at the backend to classify the data, process them and confirm information is sent out back to the device for decision making. If the data set is massive, how could an analyst handle the influx? Driverless cars, for instance, have to make rapid choices when on autopilot and relying on humans is completely out of the picture. That's where Machine Learning comes to play with its.

Consumer versus Industry Side AI representation:

The value and the promises of both AI and IoT are being comprehended because of one another. **Machine learning (ML) and Deep learning (DL)** have led to huge leaps for AI in recent years. As mentioned above, machine learning and deep learning require enormous amounts of data to work, and this data is being composed by the billions of sensors that are enduring to come online in the Internet of Things. **IoT makes better AI**. Enlightening AI will also drive embracing of the Internet of Things, creating a worthy cycle in which both areas will hasten drastically. That's because AI **makes IoT useful**.

- **On the industrial side**, AI can be applied to forecast when machines will need maintenance or examine manufacturing processes to make big efficiency gains, saving millions of dollars.
- **On the consumer side**, rather than having to acclimatise to technology, technology can adapt to us. Instead of clicking, typing, and searching, we can merely ask a machine for what we need. We might ask for information like the weather or for an accomplishment like preparing the house for bedtime (turning down the

thermostat, locking the doors, turning off the lights, etc.). (Remember Hive Home in the UK using IoT platform to deliver this? Or even Amazon Alexa using ASR technology to make this happen).

AI and IoT are Indivisibly Intertwined:

- Relationship between AI and IoT is much like the joining between the human brain and body.
- Our bodies collect sensory input such as sight, sound, and touch. Our brains take that data and makes intellect of it, turning light into identifiable objects and turning sounds into reasonable speech. Our brains then make conclusions, sending signals back out to the body to command movements like picking up an object or speaking.
- All of the connected sensors that make up the Internet of Things are like our bodies; they afford the raw data of what's going on in the world. Artificial intelligence is like our brain, making sense of that data and determining what actions to perform. And the connected devices of IoT are again like our bodies or body parts, carrying out physical actions or interconnecting to others.

AI & IOT Advances at The Core of Innovation

Back in 2018, when Sophia, a humanoid robot, sang a duet with Jimmy Fallon at his show, the whole world was in awe of how the robotic intelligence showcased human emotions while performing a song [3]. The creator, David Hanson not just invented artificial intelligence that reflects human intelligence, but also enabled it to show human emotions.

This is a major breakthrough in the technology world and is just a teaser of what is about to become a reality. While science fiction, a popular genre of movies and books, has already elaborated about how AI is soon to be at the forefront of tech innovations; the future holds a lot more. The endless possibilities that this technology is guaranteed to turn into reality are fascinating and have opened up a whole new subject of debate for techies.

Artificial intelligence in itself is a giant leap in the technology world. Combining it with inventions like IoT has further aided in utilizing the full potential of both technologies. Whilst IoT connects two or more physical objects, actuators, sensors, platforms, and networks to enable data transmission for varied applications; AI is capable of examining even critical information with ease to offer valuable insights that can aid in making highly informed and productive decisions.

4.5 Strategic Trends 2018 and Beyond

As shown in the Fig. 4.3, the technology trends change according to the following defined applications and its requirements.

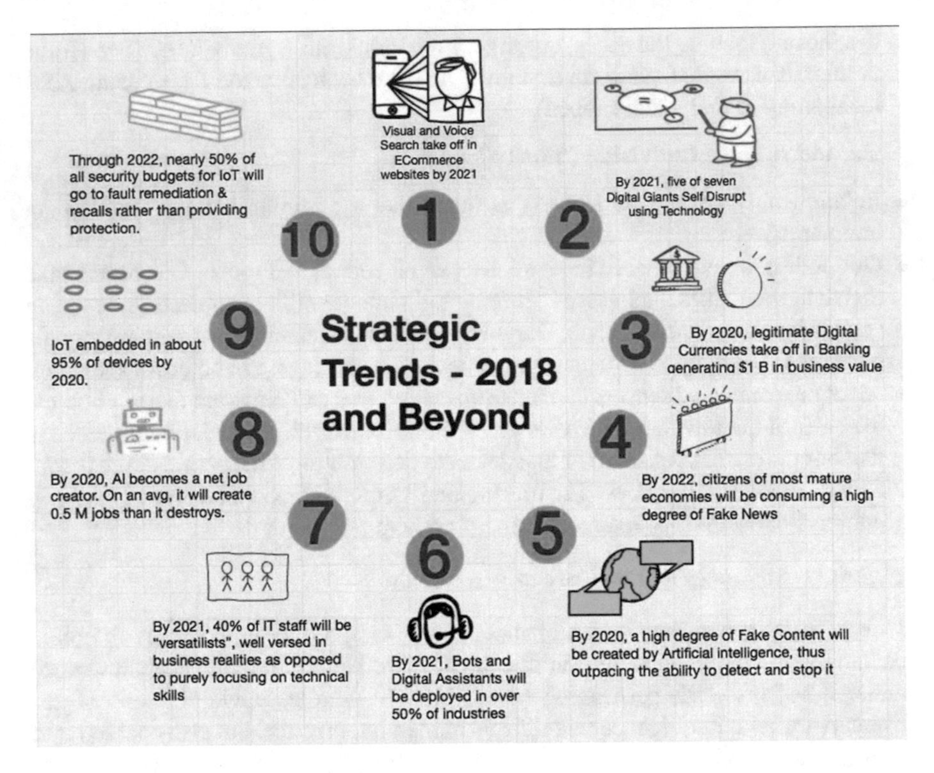

Fig. 4.3 Trends in 2018 and beyond

Transforming Business Operations

In the past decade, the dynamics of the business world have evolved rapidly with the upsurge of competition; majorly due to start-ups expanding to new verticals and successfully capturing huge market shares. In this scenario, to steal a lead, businesses are rooting for innovations that enable smart data exchange; while examining the same and offering observations that can be further deployed for making amendments that can be remunerative.

Adeptly meeting this requirement, AI and IoT together enhance the operational efficiency of business processes, maintaining the highest order of efficiency through and through. Merging **Artificial Intelligence with the Internet of Things** is a maturity progressing that is playing a key role in the growing operational efficiency of commercial organizations across diverse business verticals. The purpose of combining both the technologies was to generate smart machines that can make analytical decisions with negligible or no human support.

Enhancing Efficiency in Manufacturing

Modern manufacturing units have evolved from dovetailed workers and mechanically operated machinery to an exceptional blend of tech innovations with little support rendered by human operators [6]. Today, the premises of a manufacturing business comprise of inventive technologies like robotic arms that are often controlled remotely; all thanks to in-built IoT sensors and AI. While the **application of artificial intelligence** and IoT in manufacturing enables precise monitoring of operations, it also aids in conscientious assessment and resolution of possible miscues.

The innovative fusion of both technologies facilitates the compilation of data and implementation of algorithms to predict operational anomalies and enable timely resolution. Gradually, via machine learning, the AIIOT (Artificial Intelligence of the Internet of Things) starts recognizing external and internal factors that lead to a particular glitch.

This information is analysed over a period of time and is deployed for required repair and adjustments. The predictive maintenance initiated via AIIOT reduces possibilities of downtime and systemic failures, thus boosting the operational efficiency of the manufacturing units whilst significantly streamlining operations.

Advancement in Healthcare and Pharmaceuticals

Just like every industry, healthcare and pharmaceuticals also generate a massive amount of data in the form of medical reports, prescriptions, patient history, medicinal formulations, and pharma manufacturing details. This data needs to be not just recorded but also analysed to administer treatment as per a patient's history and manufacture medicinal formulations [7]. There are many other tasks related to data accumulation in the healthcare and pharmaceutical industry, which is why the **application of artificial intelligence** and IoT seems pragmatic.

The advent of this technology has aided in not just administering the best treatment to patients but has also helped professionals understand medical cases in a better way [9]. Apart from all this, the fusion of these technologies is also deployed to predict fitness and health issues. AIIOT powered devices like Fitbit offer valuable information regarding the activity and fitness levels of individuals; thus, generating reports regarding the possibilities of diseases and preventative measures, helping users stay hale and hearty for a long time.

AI & IoT Innovation Congregates Education

Education is the foundation of a healthy society that is capable of birthing professionals and individuals who can succeed in various domains, contributing to socioeconomic development. This is the reason why innovation in teaching methods is crucial, considering the rapid transformation that we, as a society, are undergoing [8].

Today, educational institutes aren't just reforming the way of imparting knowledge to pupils but are also advancing to the educational system of digital classrooms. In this aspect, technologies like artificial intelligence and IoT are deemed to be immensely helpful. Schools across metropolitans like Beijing are integrating technologies like AIIOT in their teaching methods to make classroom sessions more engaging and productive.

The AIIOT Advantage

As AI brings crucial machine learning and intelligence to IoT, the possibility to reap maximum benefits of both the technologies expands [8]. The deluge of leverages that AIIOT offers as a technological fusion has made it a buzzword that is impacting numerous industries.

Let's examine some major implications that AIIOT brings to the table in order to attain a better understanding of this innovation and its advantages.

- As you must be aware, artificial intelligence application enables processes like data analysis, image recognition, and language processing. These processes, with the aid of the Internet of Things, can facilitate logical reasoning and decision making, at a much faster pace than the human workforce can possibly accomplish. This certainly empowers organizations to enhance their operational competency and make decisions that are highly cogent, with almost negligible scope for errors.
- The interdependence of both technologies results in precise, prescriptive, and predictive analytics that can be deployed to modify operations, products, and services. This turns out to be a transformative impact that works in favour of the growth and development of organizations.
- In the piles of data received via the interconnected IoT devices, AI integration helps in segregating time-sensitive information that needs to be processed expeditiously. This further aids in the reduction of pending tasks that are urgent and are capable of immensely affecting the overall operational competence.

There are a lot more advantages that have contributed to reinforcing the position of AIIOT in data management and processing. As we investigate into this topic, it is crucial to realize how transformative this technological fusion is for not just industrial verticals and corporates but also our day to day lives. In fact, we aren't far from the day when AIIOT will become a part of our routine. Interconnected devices will run our day, taught by machine learning and implementations of AI to boost our productiveness and even health.

Recently, a prominent medical device company launched a mobile application to make pacemaker data more accessible for the patient as well as their treatment providers. This is just one example of how AIIOT can bring about monumental changes in the way we live.

Can you envision how ground-breaking it can be for the treatment of patients with chronic heart conditions? It can boost the survival rate of patients and can even help them lead a highly healthy life by consistently offering valuable insights to bring changes in lifestyle as well as medication. This tech fusion sure is about to evolve our world in endless ways, as demonstrated by its growing involvement in different walks of life. AIIOT shall soon be at the core of innovations that shall surround our day to day schedule, from our home to our workplace.

4.6 An Analysis of Machine Learning and IoT in Smart Transportation System

Over the last decade, applications based on mobile devices, sensors, and actuators have become smarter, empowering the communication among devices and the accomplishment of more complex tasks. In 2008 the number of connected devices surpassed the global population [1] and the number keeps increasing exponentially until today. Smart phones, embedded systems, wireless sensors, and almost every electronic device are associated to a local network or the internet, leading to the era of the Internet of Things (IoT). With the number of devices increasing, the amount of data collected by those devices is increasing as well. New applications emerge that analyse the collected data to make meaningful correlations and achievable decisions, leading to Artificial Intelligence (AI) via Machine Learning (ML) algorithms.

4.6.1 Internet of Things

In IoT terms, every connected device is deliberated a thing. Things usually consist of physical sensors, actuators, and an embedded system with a microprocessor. Things need to communicate with each other, creating the need for Machine-to-Machine (M2M) communication. The communication can be short-range using wireless technologies such as Wi-Fi, Bluetooth, and ZigBee, or wide-range using mobile networks such as WiMAX, LoRa, Sigfox, CAT M1, NB-IoT, GSM, GPRS, 3G, 4G, LTE, and 5G [2]. Due to the enormous usage of IoT devices in all kinds of everyday life applications, it is essential to keep the cost of IoT devices low. Additionally, IoT devices should be proficient to handle basic tasks like the data collection, M2M communication, and even some pre-processing of the data depending on the application. Thus, it is required to find a balance among cost, processing power, and energy consumption when designing or selecting an IoT device. IoT is also tightly attached to "big data", since IoT devices continuously collect and exchange a great amount of data. So, an IoT infrastructure usually implements methods to handle, store, and analyze big data [3]. It has become a common practice in IoT infrastructures, to use an IoT platform such as Kaa, Thingsboard, DeviceHive, Thingspeak, or Mainflux in order to support the M2M communication, using protocols like MQTT, AMQP, STOMP, CoAP, XMPP, and HTTP [4]. Additionally, IoT platforms offer monitoring capabilities, node management, data storing and analysing, data driven configurable rules, etc. Depending on the application, it is sometimes essential that some data processing takes place in the IoT devices instead of some centralized node as it happens in the "cloud computing" infrastructure. So, as the processing partially moves to the end network elements, a new computing model is introduced, called "edge computing" [5]. However, since those devices are most of the times low-end devices, they may not be suitable to handle intense processing tasks. As a result, there is a need for an intermediate node, with sufficient resources, able to handle advanced processing tasks, physically located close to the end network elements, in order to minimize the overload caused by massive sending of all the data to some central cloud nodes. The solution came with the introduction of the "Fog nodes" [6]. Fog nodes help IoT devices with big data handling by providing storage, computing, and networking services. Finally, the data are stored in cloud servers, where they are available for advanced analysis using a variety of ML techniques and sharing among other devices, leading to the creation of modern added value smart applications. IoT applications have already emerged in many aspects the so called, smart city. We could group the most important applications in the following categories [7]:

- **Smart Homes**: This category includes old-fashioned home devices, such as fridges, washing machines, or light bulbs, that have been developed and are able to communicate with each other or with authorized users via internet, offering a better observing and management of the devices as well as energy consumption optimization. Apart from the traditional devices, new technologies spread, affording smart home assistants, smart door locks, etc.
- **Health-care assistance**: New devices have been developed in order to improve a patient's well-being. Plasters with wireless sensors can monitor a wound's state and report the data to the doctor without the need for their physical presence. Other sensors in the form of wearable devices or small implants, can track and report a wide variety of measurements, such as heart rate, blood oxygen level, blood sugar level, or temperature.
- **Smart Transportation**: Using sensors entrenched to the vehicles, or mobile devices and devices installed in the city, it is possible to offer enhanced route suggestions, easy parking reservations, economic street lighting, telematics for public means of transportation, accident prevention, and autonomous driving.
- **Environmental Conditions Monitoring**: Wireless sensors dispersed in the city make the perfect infrastructure for a extensive variety of environmental conditions monitoring. Barometers, humidity sensors, or ultrasonic wind sensors can assist to create advanced weather stations. Moreover, smart sensors can monitor the air quality and water pollution levels across the city.
- **Logistics and Supply Chain Management**: With the use of smart RFID tags, a product can be easily tracked from the production to the store, reducing cost and time considerably. In addition, smart packaging can offer features such as brand protection, quality assurance, and client personalization.
- **Security and Surveillance Systems**: Smart cameras can obtain video input across the streets. With real-time visual object recognition, smart security systems can identify suspects or prevent hazardous situations.

As considered so far, although there is a lot to be done in terms of standardization when it comes to IoT infrastructure and technologies, Fig. 4.4 can accurately describe the key elements of the infrastructure as they have been used in the majority of the applications.

Figure 4.4 is structured by separating the infrastructure key elements in totalled blocks. Each block portrays a illustrative image of the described element, and arrows are linking the images with numbers, indicating how each element intersects with

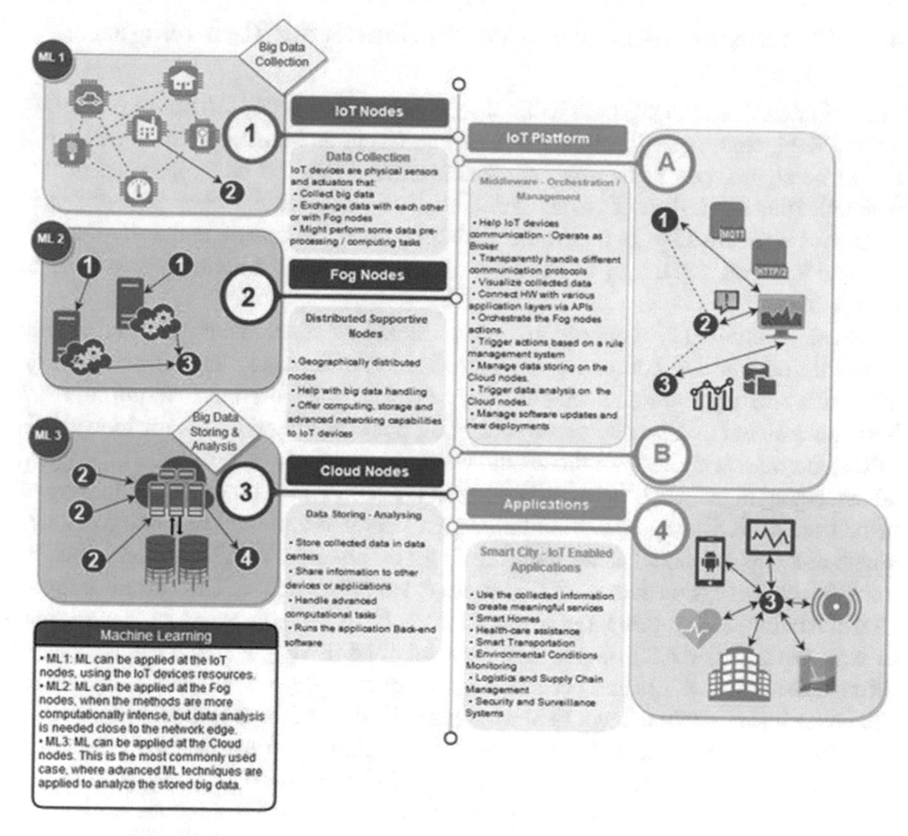

Fig. 4.4 Key elements of the Internet of Things (IoT) infrastructure

the other blocks. Additionally, text blocks are included, giving the most important aspects of each key element in bullets. The IoT infrastructure consists of: (1) IoT nodes, namely the IoT devices (sensors and actuators) at the edge of the network. (2) The Fog nodes, servers that help IoT devices by providing computing, storage, and advanced networking capabilities. (3) Cloud nodes that contain of data centers, which handle data storing, computationally intense data analysis using ML techniques, data sharing, etc. (4) IoT applications that use the collected and analysed information to generate services for the end user. Points (A) and (B) symbolize the IoT platforms, which have the orchestrator's role and support the elements throughout the infrastructure in numerous ways as described in the figure. Moreover, Fig. 4.4, presents the role of ML in the IoT infrastructure. ML techniques can be pragmatic at the IoT nodes, the fog nodes, or the cloud nodes, contingent on the application needs.

4.7 Conclusion and Future World-Shattering Technologies

The common thread in all these technologies is DATA. DATA is the one connecting all of these technologies. Data has to be learnt, has to be transmitted back and forth, has to be stored, has to be analysed, data patterns have to be explored and data authentication and ledger has to be maintained. These different technologies come at numerous different steps in this value chain. Not essential that all technologies have to come together, some may and some may not, based on the actual prerequisite of the use case.

These revolutionary technologies are expected to amendment our routine life; just as introduction of computers did, introduction of internet did, and introduction of mobile/smart phones devices did. A considerate collaboration is required, from both government and private sectors; to bring out cost-effectiveness for industrialisation, alternative choices for those who would be affected and further technological advancement to make it a sustained reality. It is not crucial yet for these technologies to be tried out with urgency. Planning/assessment can certainly be done to check, whether any real-benefit can be achieved. If there is benefit, then investment must be made in line with "considerate collaboration" we talked above. The current impression of it being reachable to large corporations only; that it required a lot of investment to get it realised; and it is going to take away lot of human engagements; must change through better awareness and perhaps economies of scale. It's happening…for sure. My favourite areas where I would like to see application of these technologies are: for revitalization of agriculture industry; to combat global warming/extreme weathers and finally in the pharmaceutical industries to find cures for deadly diseases.

An analysis on Machine Learning and Internet of Things techniques subjugated for smart transportation applications has been presented. This study highlighted the circumstance that an extensive variety of Machine Learning algorithms has been proposed and evaluated for Smart Transportation applications, indicating that the type and scale of IoT data in these applications is ideal for ML exploitation. On the other hand, given the current applications and infrastructure regarding IoT and ML, a comparatively smaller ML exposure for the smart lighting systems and parking applications is detected. Therefore, there is a definite need for supplementary exposure in those areas, from the ML perspective, in the future.

Concerning the IoT tactics for their application categories, route optimization, parking, and accident prevention/detection have proven to be the most prevalent among them. Considering the difficulties that the smart transportation applications address, some common points of interest have been identified from this evaluation. System defined with the following operations namely human safety, environmental precautions, transport finance and time saving.

Moreover, the great progress which has already been attained in the field of smart transportation with the help of IoT and ML became apparent, while an even better advancement in this topic is expected in the upcoming years. As the number of IoT devices rises, the data diversity and volume scale up, therefore, ML can produce many evocative applications.

References

1. https://channels.theinnovationenterprise.com/articles/aren-t-the-iot-big-data-and-machine-learning-the-same
2. Balaji, S., Harold Robinson, Y., Golden Julie, E.: GBMS: A new centralized graph based mirror system approach to prevent evaders for data handling with arithmetic coding in wireless sensor networks. Ingénierie des Systèmes d'Information. **24**(5), 481–490 (2019)
3. Harold Robinson, Y., Rajaram, M.: Energy-aware multipath routing scheme based on particle swarm optimization in mobile ad hoc networks. Sci. World J., 1–9 (2015)
4. https://analyticstraining.com/how-to-apply-machine-learning-algorithms-to-iot-data/
5. Harold Robinson, Y., Rajaram, M.: A memory aided broadcast mechanism with fuzzy classification on a device-to-device mobile ad hoc network. Wirel. Pers. Commun. **90**(2), 769–791 (2016)
6. https://www.mdpi.com/1999-5903/11/4/94/htm
7. Harold Robinson, Y., Golden Julie, E., Balaji, S., Ayyasamy, A.: Energy aware clustering scheme in wireless sensor network using neuro-fuzzy approach. Wirel. Pers. Commun. **95**(2), 703–721 (2017)
8. Corrin, L., Kennedy, G., de Barba, P.: Asking the Right Questions of Big Data in Education (2017). https://pursuit.unimelb.edu.au/articles/asking-the-right-questions-of-big-data-in-education
9. Marr, B.: How Big Data Is Changing Healthcare. Forbes/Tech (2015). https://www.forbes.com/sites/bernardmarr/2015/04/21/how-big-data-is-changing-healthcare/#6643a6972873
10. Smyth, D.: Why blockchain? What can it do for big data? (2016). http://bigdata-madesimple.com/why-blockchain-what-can-it-do-for-big-data-2/

Chapter 5
A Security and Confidentiality Survey in Wireless Internet of Things (IoT)

Arpan Garg, Nitin Mittal and Diksha

Abstract Internet of Things (IoT) is regarded one of the evolving technologies that offers many vertical industries excellent possibilities. The IoT paradigm paves the way for a world that connects many of our every object and communicates with their environments for information collection and task automation. Such a vision requires data privacy, safety, assault robustness, simple deployment, and self-maintenance. Wireless IoT is a successful area that continually collects, exchanges and stores big amounts of data and information. While these data and information are used in wireless IoT for excellent reasons, major dangers emerge as attackers seek to take advantage of the merits of this latest technology for their own benefit. IoT pervasive growth is noticeable worldwide. Devices are readily exposed in such a resource-constrained setting. This paper presents the classic wireless IoT application scenarios, along with associated models of safety and privacy assault. This paper further presents recent advances in Wireless IoT privacy conservation mechanisms along with classification of the application scenarios. In this paper, open problems and future research directions are addressed for the implementation situations in wireless IoT.

Keywords Cyber security · IoT · IoE · IoS · IoV · Cyber attack · Sybil attack · Malicious attack · Cryptography

5.1 Introduction

The IoT is a network of physical objects including cars, meters, equipment, and other electronically integrated products, software, sensor and actuators. All IoT

A. Garg · N. Mittal (✉) · Diksha
Electronics and Communication Engineering, Chandigarh University, Punjab 140413, India
e-mail: mittal.nitin84@gmail.com

A. Garg
e-mail: arpan04garg@gmail.com

Diksha
e-mail: 13dishathakur@gmail.com

V. E. Balas et al. (eds.), *Internet of Things and Big Data Applications*, Intelligent Systems Reference Library 180, https://doi.org/10.1007/978-3-030-39119-5_5

objects have an internet connection communicate and exchange information. IoT can implement a variety of attractive applications, such as smart recognition, location identification, smart monitoring, oversight and management, and so on based on communication and information exchange. IoT-related devices and items were projected to rise to around 30 billion by 2020, or nearly five devices for every person worldwide. IoT will achieve $7.1 trillion in worldwide market value in 2020. The exponential growth of IoT is very evident and it is the motive for the digitization of the logistics industry and lays the foundations for Industry 4.0. Wireless IoT can connect integrated wireless phones to each other. Generally, this type of device has restricted capacity to compute, process, transmit and store and is energy restricted as well. As a consequence, almost every domain can find wireless IoT applications properly. Smart grids, e-health care, intelligent homes, intelligent transport and intelligent cities form part of the typical situations of wireless IoT. They have same essence of wireless IoT despite their various features in various application fields.

This implies that each application can be seen as a network of wireless IoT devices interconnected. Intelligent decisions can be produced on the grounds of timely data collection and communication, to promote and enhance people's quality of life. With the fast growth of wireless communications systems over latest years, a host of integrated and mobile devices in multiple situations have been widely available, with billions of interconnected wireless stuffs distributed across distant regions and big quantities of delicate and critical information generated, processed and exchanged to support innovative services.

Various things generate abundant data transmission in the wireless IoT setting. This information may include the conduct, tastes, preferences and critical private information of the user. Devices and services are subjected to enormous risks in such a vibrant and distributed setting as wireless IoT, which can jeopardize their valued information and eventually the personal identity of their customers. An adversary may misuse customer privacy information or damage it by tapping, leaking, changing and destroying it.

Of course, the insufficiency of WI-FI in addressing various assaults. The failure to implement efficient data protection policies continue to be significant barriers to ensure that Wi-Fi succeeds in gaining participants confidence as it addresses most safety issues indirectly. For wireless IoT systems where intelligent objects can be independent, independent entities with their own identity are therefore extremely expected to be effective. Researchers created various privacy-reserving authentication protocols in Wireless IoT in order to attain these goals. This paper deals with wireless IoT privacy authentication, which is structured in the following way. Sections present scenarios for Wireless IoT applications and discuss the associated models of privacy threats. Then provide a short summary of the data protection authentication policies cited. In several wireless IoT environments, the recent advances in privacy protection authentication protocols are detailed.

5.2 Wireless IoT Applications

Applications from scientific literature for Wireless IoT and Smart City environments in the last few years received considerable attention. Smart Cities can provide their digital citizens, for instance for emergencies and health care [1] as shown in Fig. 5.1.

A variety of research topics, including event forecasts [2], routing protocols [3] for WSN, fusion of multi-sensor information [4], company model and profit maximization [5], ontology [6], service models [7], quality of service [8], and even sophisticated concepts for the emphasis on raw data processing and dissemination of information such as information quality (QoI) [9] and Value-of-Information [10] are the part of wireless applications. A variety of application-specific alternatives for diagnostic problems [11, 12], environmental surveillance [13, 14] and social concern [15] were also created by researchers. However, the design and growth of wireless IoT applications in spare Smart City settings still reflects a major challenge, requiring both the architectural and the Communication paradigms to develop fresh designs and paradigms. Fog Computing is one of the most exciting and relevant study fields promoting the creation and administration of Wireless IoT applications in Smart City settings from an architectural perspective. The Stack4Things platform is for example an IoT application for the management of sensors and actuators, grouping them together, making it easy for them to interact and allowing their behavior to be specialized. The Stack4Things is the cyber-physical system with the CSFV functions (CPSFV) [16] as shown in Fig. 5.2.

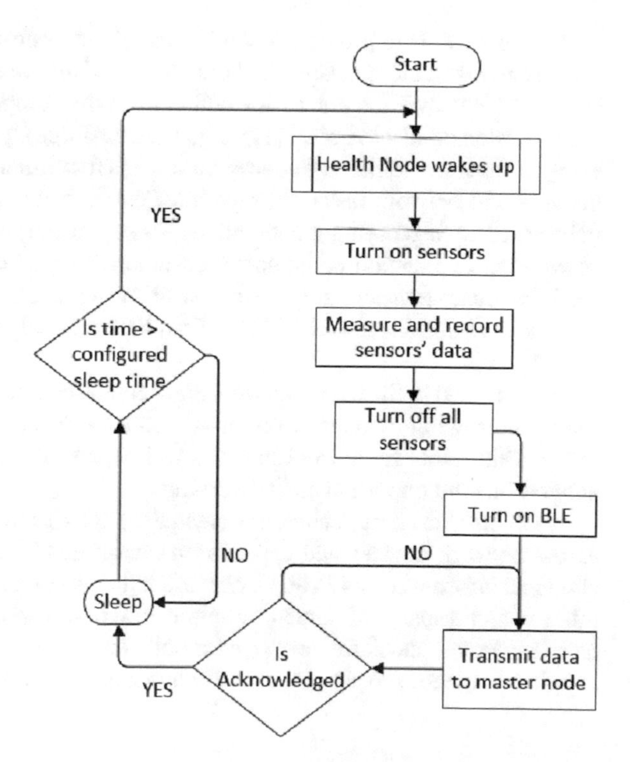

Fig. 5.1 For emergencies and health care [1]

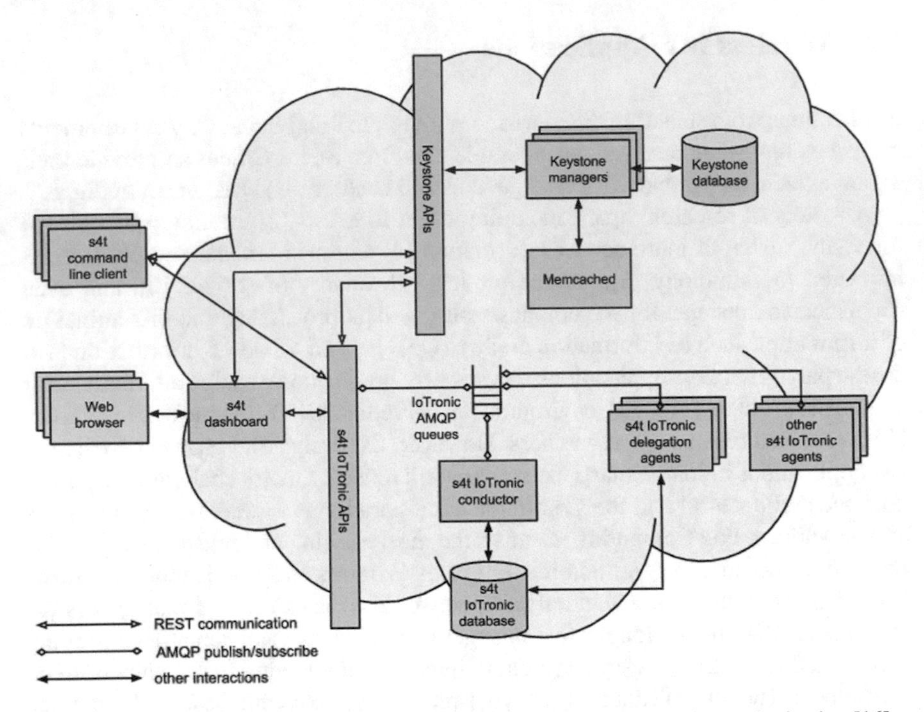

Fig. 5.2 Cloud-side-Stack4Things-architecture with focus on authentication-authorization [16]

The last task is to provide context to the cloud-controlled injection of code into every sensor/actuator managed in the shape of plugins on the Stack4Things platform. In order to leverage the processing authority of the components in a network infrastructure, Manzalini and Crespi [17] suggested the Edge Operating System (EOS) software architecture, reinforced concepts and tools for virtualization of software-defined network and network tasks. Information Centric Networking is another promising field of research presenting alternatives for supporting wireless and heterogeneous networking application communication in small smart cities, which also facilitate the implementation and administration of Wi-Fi applications. Amadeo et al., for example, submitted an ICN MANETS architecture called CHANET [18] for IEEE 802.11.

Mendes [19] defined an opportunistic, content-centered architecture that benefits from the growing amount of omnibus technologies accessible for content sharing today. This ICON framework involves techniques derived from both data-centered networking and opportunistic networking.

ICON uses caching techniques created by ICN for sharing and placing content across network devices and depends on opportunistic approaches based on social and local information and knowledge about the use of application data for the transmission and transfer of content to interesting nodes. Finally, Delmastro et al. [20] have suggested specially developed mobile social networks in order to encourage people to engage actively in the development and sharing of quality of lifetime data.

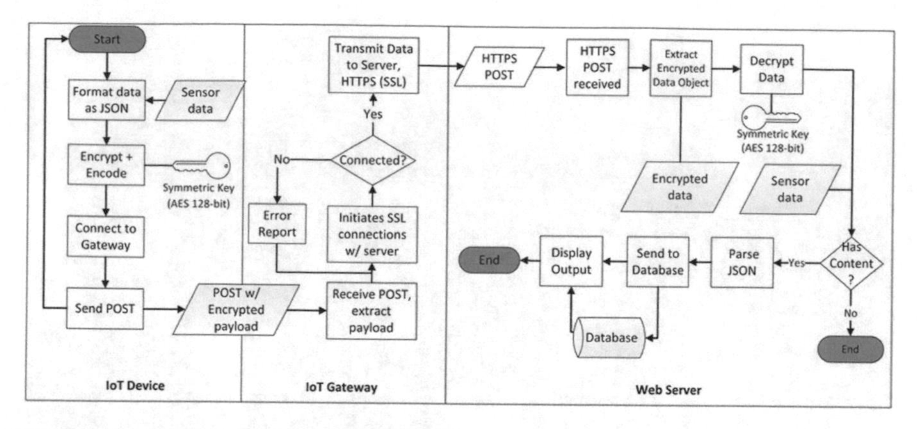

Fig. 5.3 Security based IoT components [23]

A significant subject on the communication level is the flexibility and dynamic management of packet dispatch in IoT settings. For example, a solution that can help to save energy when sending packets over reliable wireless networks Network Code and Power Control based Route (NCPCR). NCPCR uses network coding for dynamic power transmission management. Chen et al. [21] proposed that cognitive radio be taken into IoT environments to provide an interference-wake flood scheme for the enhancement of information projections between nodes to achieve a vibrant and efficient variety. Various IoT technologies were defined and contrasted by Al-Turjman [22] in terms of the energy consumption, expenses and delays. The newly suggested routing of the Multipath Disruption-Tolerant Approach (MDTA) allowed a reduction in message delivery delays, while decreasing the use of roadside fixed nodes in sensor networks. Finally, ant-based Efficient IoT Communications (EICAntS) enable the network performance to be improved with respect to nodes life and energy use by introducing an uninspired IoT-based routing protocol [23] as shown in Fig. 5.3.

Wireless IoT applications can benefit from leadership choices based on dynamically collected context data at the co-ordination stage. Pilloni et al. [24], for example, suggested a concerted strategy designed to maximize the life span of node groups to meet the QoI criteria. To this aim, nodes coordinate one another using a consensus strategy to optimize assignment of duties specifically considering the lifespan of nodes, storage capacities, processor abilities and accessible bandwidth. Cao et al. [25] provided more specifically state-of-the-art methods of cooperation in the industry. In specific, it introduces a few efforts to reach consensus in distributed situations among several actors. Caraguay et al. [26] suggested a better network management solution, considering new networking paradigms such as SDN and NFV. The suggested solution SELFNET enables big quantities of controlled information collected from heterogeneous sources to be readily queried. In order to provide improved data and enhance the quality of service (QoS) network, SELFENT data can also be linked through appropriate coordination and implementation of the right measures as shown in Fig. 5.4. While cutting-edge solutions provide interesting insights into the problems of wireless multi-hop networks, note that they are silos-based and do not offer a

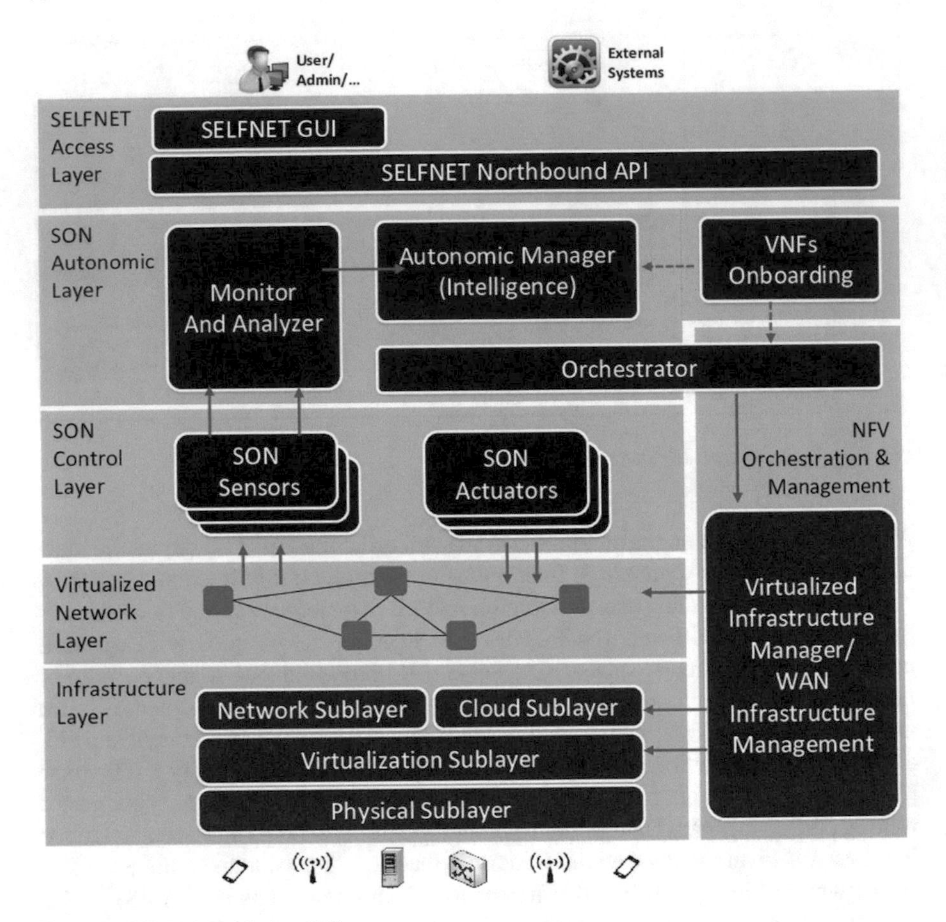

Fig. 5.4 SELFNET solution [26]

full architectural solution covering subjects in separate layers. Delmastro et al. [20] have also aimed at citizens cooperation in cities of intelligent communication based on opportunistic sensing, their research gives information on the implementation created; moreover, it reports outcomes of performance primarily on the user's quality of experience and on easy subnet working (without assistance for inter-subnet routes).

Sotres et al. [27] concentrated primarily on the many practical elements of the deployment of large-scale Smart City facilities and overlooked problems relating to the growth and implementation, in addition to technology, of new applications. The development of wise overall methodologies (and related prototyping instruments) has been considered as relatively neglected in order to promote the growth of wireless IoT apps for communications, coordination and administration. The remainder of the article offers a solution to readily set up and manage wireless monitoring and control systems in smart cities. Our proposition offers alternatives for scientists and professionals in the sector (thrown-out subnet configuration process and the Multi free Packet Shipping option), from the Communication Layer to the coordination of nodes involved in the same wireless IoT implementation.

5.3 Different Security Attacks in IoT

According to many scientists, IoT technology works on three levels of sensing, networking and execution, as illustrated in Fig. 5.5 [28]. Perception Layer contains various types of detector data, such as RFID, bar codes, and other networks. This layer is intended to use sensors to collect environmental information and then transmit it to a network layer. The goal of the network layer is to communicate data collected from the receiving level through the internet, mobile network or any other trusted network to any information treatment system. IoT's goal is to produce a smart application layer atmosphere. In terms of complexity, heterogeneity and a big number of interconnected assets, IoT safety poses a significant challenge. Using a malicious program and encryption strategy, i.e. damaging or manipulating a node, i.e. a physical vulnerability, or misusing a routing protocol and other network-related protocol, the adversary may be able to connect the IoT system. Classify IoT attack as a physical attack, network attack, computer attack, and encryption attack into four categories. We considered one category of attack to be the most dangerous of all the violence in this category.

A physical assault on the injection node is a physical attack. Not only the services are terminated, but the information changes are also halted. The Sinkhole attack is the most dangerous type of network attack. Not only does it attract all traffic to

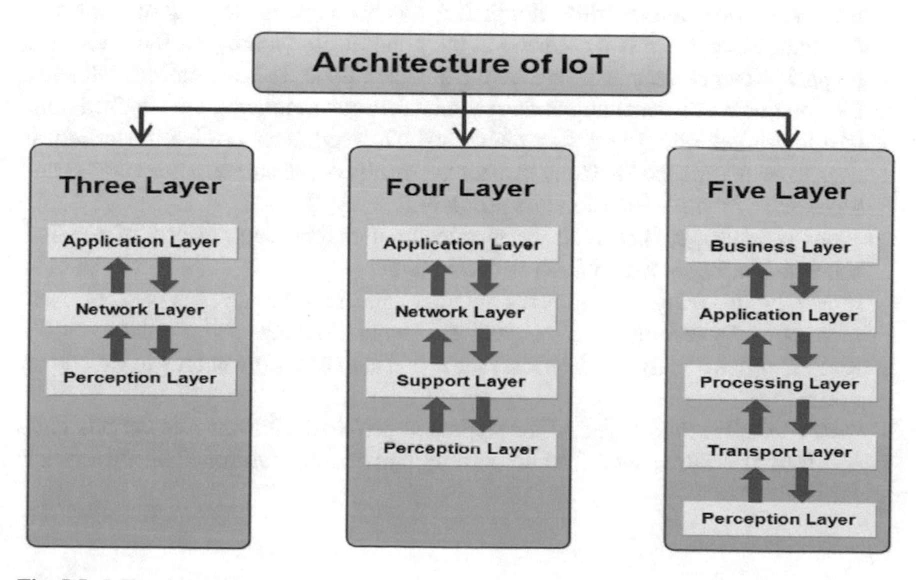

Fig. 5.5 IoT architecture

the base station, but other threats can also occur such as selectively forwarding, changing or dropping packets. The worm attack we choose from is the most unsafe from the software attack. Worms may be the most damaging and destructive of internet malware. It is the self-replicating program that harms the computer with safety troubles in networking software and hardware.

Side channel attack is the hardest to manage from the encryption attack. It's very hard to identify because the attacker utilizes the data on the side channel to conduct the attack.

5.3.1 Physical Attack

Physical assaults focus on the system's hardware devices.

1. *Node Tampering*. Physically modifies the compromised node in this attacker and can acquire delicate data such as encryption key [28].
2. *RF Interference on RFIDs*. Radio frequency signals are sent by noise, the attacker conducts Denial of service assault. These signals are used to communicate with RFID [29, 30].
3. *Node Jamming in WSNs*. The intruder can interfere with wireless interaction by using jammer. It leads a service attack to be denied [28].
4. *Malicious Node Injection*. The attacker injects a malicious node physically between two or more nodes in this attack. The Data that transfers the wrong information to the other nodes are then modified. The assailant uses different nodes to inject malicious nodes [31]. The opponent first inserts a node replica. Other malicious nodes are inserted after that. Both nodes operate to perform the attack together. Thus, there is a collision at the node of the victim. For these reasons, no packet can be received/send by the attacked node. It can therefore influence the conclusion of watchdog nodes by mistakenly announcing the attacked node (the legitimate node) as acting maliciously. Surveillance verification system is used to avoid this attack. It can inspect the results of the surveillance node(s) and recognize any malicious conduct properly.
5. *Physical damage*. The attacker physically damages components of the IoT scheme, leading in a denial of service attack.
6. *Social Engineering*. The attacker interacts and manipulates IoT system users physically. To accomplish his objectives, the attacker gets delicate data.
7. *Sleep Deprivation Attack*. The attacker's goal is to use more energy to shut down nodes [32].
8. *Injection of malicious code*: The opponent physically brings a malicious code into the IoT scheme node. The IoT scheme can be fully controlled by the attacker [32].

5.3.2 *Network Attack*

The threat to wireless networks must first be acknowledged in order to attain basic security goals. Attacks on wireless networks are goal-oriented, performer-oriented and layer-oriented assault [33, 34].

1. Goal-oriented assault are active and passive attacks [1]. Active attacks happen in situations where an assailant is using its data to interrupt network operation. For example, DoS, wormhole, black hole as shown in Fig. 5.6, spoofing and modifications are included. Passive attacks happen when a malicious user monitors sensitive network data in a way that does not interfere with the network activities in other attacks. These attacks reveal sensitive data or records from unknown network users [34–36].
2. Outside or performer-orientated attacks are assaults oriented towards performers. External attacks enable monitoring of data transmission responsibilities and the injection of false data into the network for the consumption of DoS assaults. There are attacks on the network if malicious nodes are legally acceptable nodes. After the network has trusted the source, the malicious node may start numerous assaults because the malicious node can prevent significant data from reaching the malicious node [33] as shown in Fig. 5.7.
3. Layer-oriented assault consist of multiple attacks targeting distinct network stack layers [33–37]. Physical layer attacks vary from node capture to jamming channels that can lead to DoS attacks [33]. By breaking communications protocols, layer attacks target connection protocols functionality such as DoS attacks and sinkhole attacks, are network layer assaults. Transport assaults by layer consume node resources through link requests overflowing them. Attacks on application layers include assaults like blat anting, bribery of information and malicious code [37, 38] as shown in Fig. 5.8.

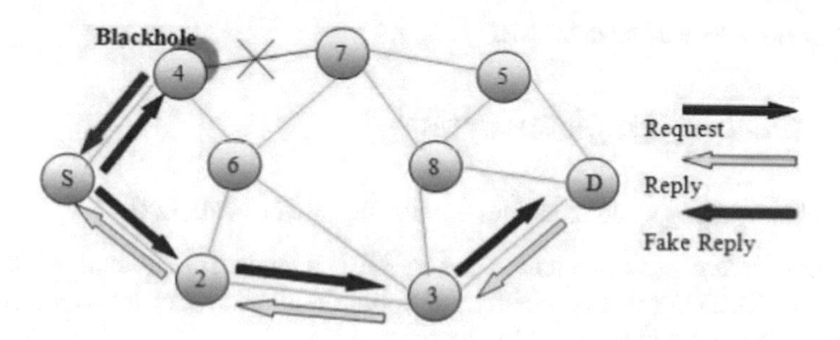

Fig. 5.6 Black hole attack

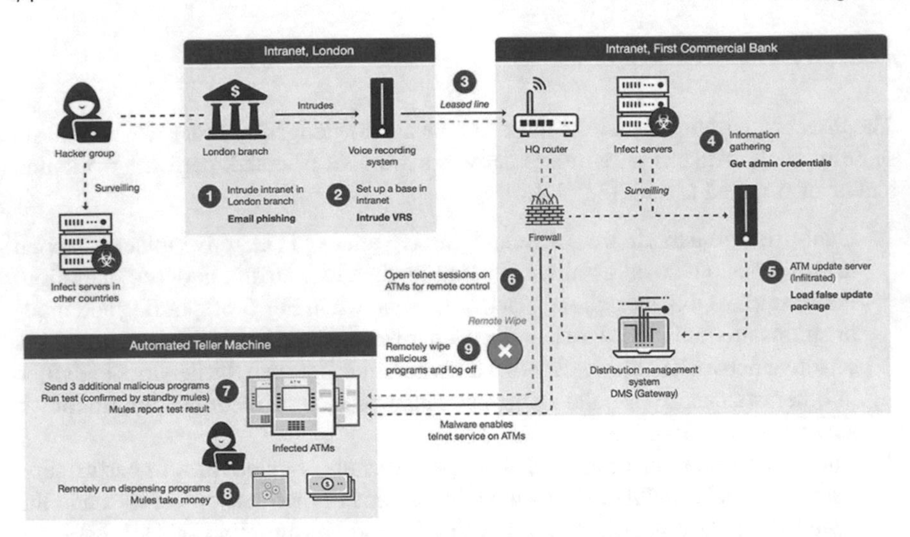

Fig. 5.7 Malicious attack [33]

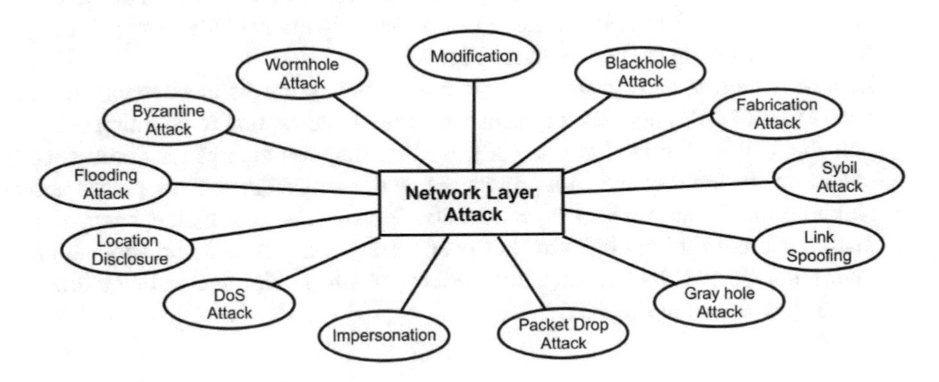

Fig. 5.8 Network layer attack in MANET

5.3.3 Attacks of Traffic Analyzes

The attacker intercepts and examines emails for network data [28].

1. *RFID spoofing*. An opponent spoofing RFID signals. It then captures the data from an RFID tag that is transferred. Spoofing attacks offer incorrect data that appears correct and accepted by the scheme [29].
2. *RFID Cloning*. In this attack, the opponent is copying the data to another RFID tag from the already existing RFID tag. The original RFID tag ID does not duplicate. Inaccurate data can be inserted, or data transmission can be controlled via cloned nodes [32].

3. *RFID Unauthorized access*. If the RFID systems do not provide the right authentication, the opponent may observe, change or delete node data [32].
4. *Sinkhole Attack*. In a sinkhole attack, an opponent compromises a node within the network and this node is used to attack. The compromised node sends wrong routing nodes that it has the minimum range route to the base station and then draws the traffic. You can then change the information and drop the packets as well.
5. *Man in the Middle Attacks*. The Internet intruder intercepts the two nodes contact. By eavesdropping the delicate data [32].
6. *Denial of Service*. The network is flooded by an attacker with big traffic so that services are not available to its expected customers [39].
7. *Routing Information Attacks*. In this attack, by spoofing, altering or sending routing data, the attacker can create the network complicated. It results in packets being allowed or dropped, incorrect information being forwarded, or the network being partitioned.
8. *Sybil Attack*. There is an attack that creates the network a lot of severe threats and it's called Sybil attack. This means that the Sybil attack on network security is an attack that undermines the reputation system through forging identities in peer-to-peer networks. In this way, attackers use numerous IDs or IP addresses to gain network control and cause a lot of misunderstanding between nodes. Examples of such systems range from email and instant messaging communication systems to cooperative content rating, recommendation and distribution systems. The system can attack on the inside or outside. External attacks, but not inner attacks, can be avoided through authentication. The mapping between identity and entity in WSN should be one by one. However, this attack violates this one-by-one mapping through the creation of several identities [40] as shown in Fig. 5.9.

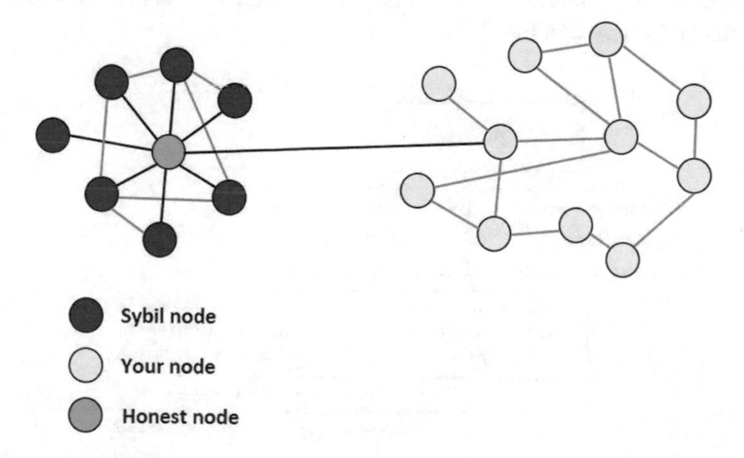

Fig. 5.9 Sybil assault

A, B, C, D are nodes of Sybil. They use any of the identities when these nodes want to connect to their neighboring nodes. This makes the network confusing and collapsing.

5.4 Encryption Attacks

These attacks are based on destroying the method of encryption and obtaining the private key.

1. *Side-channel Attacks*. The attacker utilizes data produced by encrypting equipment from the side channel. It is neither the plaintext nor the text of the cipher, it contains power information, the time required for the operation, the frequency of faults, etc. as shown in Fig. 5.10. This data is used by Attacker to identify the encryption key.

Various types of side channel attacks are present, including timing, simple and differential power analysis and differential failure analysis attacks [41] as shown in Fig. 5.11.

Timing attacks rely on how long it takes for operations to be performed. It offers information on the secret key. Fixed Diffie-Hellman exponents, RSA factor keys, and other cryptosystems can be discovered by an attacker using this information. In distinct times, cryptosystems process various inputs. RAM cache hits, processor instructions running in non-fixed moment, etc. due to branching and conditional statements.

1. *Cryptography*: Original Greek cryptography implies "secret writing." We use the term, however, to refer to transforming the message to make it safe and immune to attacks as shown in Fig. 5.12.

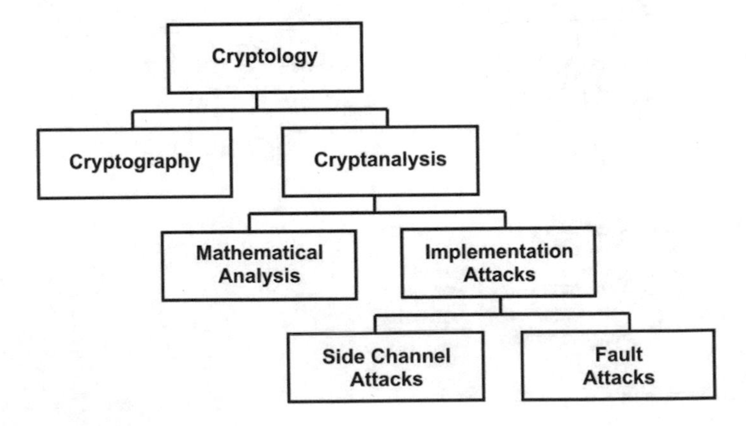

Fig. 5.10 Cryptology attacks

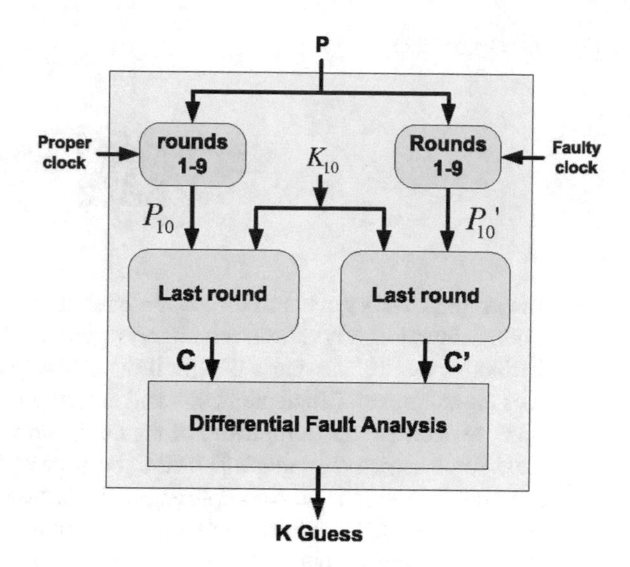

Fig. 5.11 Fault analysis attacks

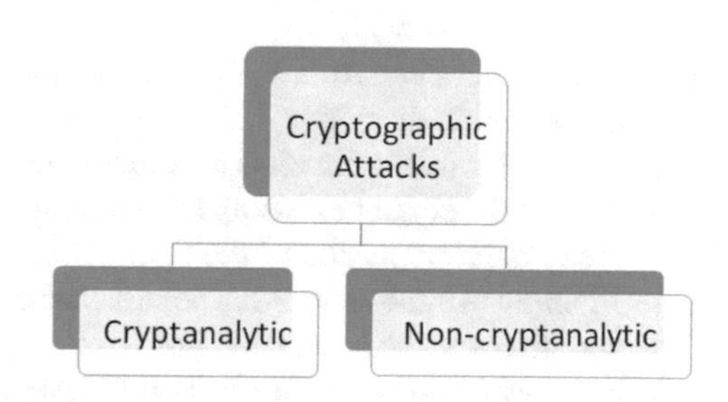

Fig. 5.12 Cryptography

Previously, cryptography only referred to the decryption and encryption of messages using the secret keys, but today it is distinct and involves three unique processes:

Symmetric-key Encipherment. In Symmetric-key Encipherment (secret-key cryptography), an entity, says Meera, can send a message to another entity, say, Ram, over an insecure channel, if somebody, says Kiran, can't comprehend the message's content by merely eavesdropping over the channel.

Asymmetric-Key Encipherment. Have the same scenario as symmetric-key encipherment in Asymmetric-Key (public important cryptography), but now there are two keys instead of one: one public main and one personal important. Meera first encrypts the email using the public key of Ram to send a guaranteed email to Ram. Ram utilizes its own private key to decrypt the message.

2. *Cryptanalytic Attacks.* These attacks are a mixture of statistical and algebraic methods designed to determine a cipher's secret key. These techniques examine the mathematical characteristics of uniform distribution cryptographic algorithms. All the cryptographic algorithms act on the distribution of the message and transform it using the key to a random-looking cipher text distribution.

Fig. 5.13 Crypto block

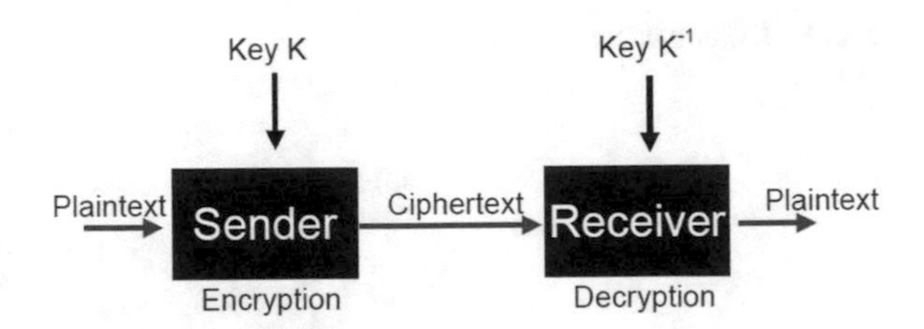

The cryptanalysis goal is to discover cipher characteristics that do not exist in random function. Therefore, the attacker guesses the key and searches for the distinctive estate. The guess is right if the property is identified otherwise next guess is attempted. Efficient attacks will embrace a method of "divide and conquer" to decrease the complexity of the conjecture key from the complexity of brute force search. An attack is said to be successful if the complexity of the guessing is lower than the complexity of the brute force. Using either plaintext or cipher text, the opponent gets the encryption key in this attack. There are distinct kinds of cryptanalysis assaults based on the methodology used [5] as shown in Fig. 5.13.

(a) *Cipher text Only Attack*. This allows the attacker to access the cipher text and to determine the appropriate plaintext [5].
(b) *Known Plaintext Attack*. The attacker understands the plaintext for certain sections of the cipher text using this technique. The purpose of this data is to decrypt the remaining portion of the cipher text [5].
(c) *Plaintext Attack selected*. The attacker can select which plaintext is encrypted and discover the key to encrypt.
(d) *Selected Cipher text Attack*. The attacker can discover the encryption key [8] by using the plaintext of the selected cipher text.

Non-Cryptanalytic Attacks. These are the attacks that do not exploit the cryptographic algorithm's mathematical weakness. However, there is still a threat to the three safety objectives, namely confidentiality, integrity, and accessibility as shown in Fig. 5.14.

3. *Man in the Middle Attacks*. The attacker intercepts the contact when two users interchange the key and gets the key [31].

Security goals. Security goals as shown in Fig. 5.15.
Confidentiality. Probably the most prevalent element of information security is confidentiality, we need to safeguard it. An organization/company must safeguard against this malicious action that threatens its information's confidentiality. In the military, concealment of sensitive information is a major concern. Hiding some data from rivals while in industry is essential.

Integrity. Information is continually changing. Like in banks, the balance of his account requires to be altered when a client withdraws or deposits cash. Integrity implies that only authorized organizations and the approved mechanism need to

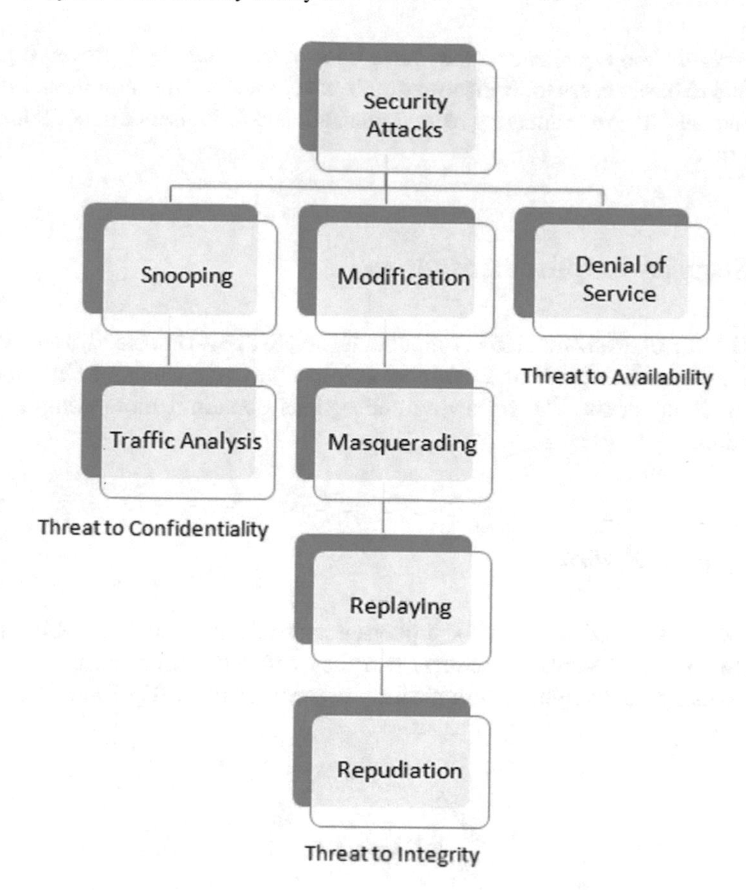

Fig. 5.14 Cryptanalytic Attacks

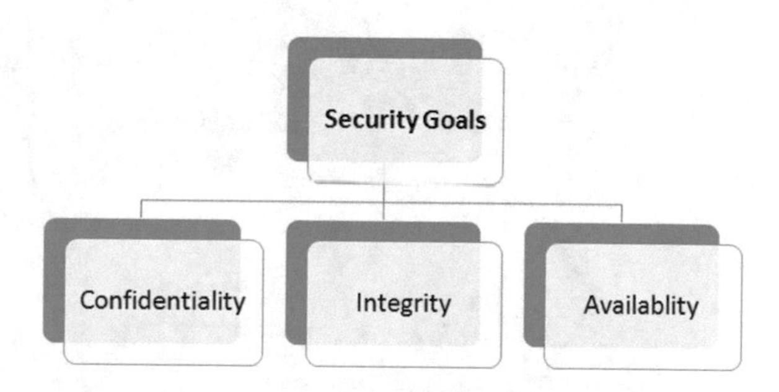

Fig. 5.15 Security goals

make the shift. Not only malware operates, but a system interruption can also generate undesirable modifications in some data that may also violate the system's integrity.

Availability. An organization may have a lot of data that the approved organizations need to have access to. If information is not accessible, it is pointless. Information unavailability is as damaging to an organization as it is absence of confidentiality or integrity.

5.5 Security Requirement for IoT

Figure 5.16 displays six main IoT components (i.e., IoT network, cloud, user, attacker, service, and platform). We find the most efficient way of checking safety demands from the components. The following subsections contain a more comprehensive description.

5.5.1 *IoT Network*

IoT network is a particular type of modern network. There are many objects in the IoT network (e.g. gateways, sensors) and they can interact on the basis of IEEE 802.15.4 using lightweight communication procedures like MQTT and CoAP.

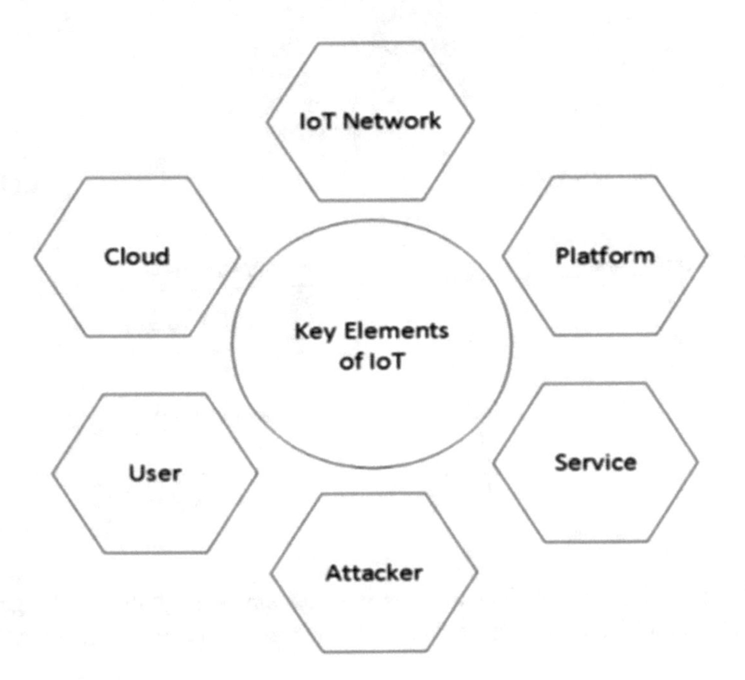

Fig. 5.16 Security requirement for IoT

Most importantly, the IoT network is not fundamentally distinct from standard networks. Most current issues (e.g. fragmentation, safety assaults) in the IoT network could therefore occur. We concentrate on the following problems in this subsection: privacy, multicasting safety and bootstrapping.

Privacy: IoT is becoming as ubiquitous as ever nearer to human existence. It can be used with anything anywhere, anytime. People will be supervised everywhere by CCTV and any sensed data will be sent to the network by sensors. In addition, there will be a diversification of kinds of information and a gradual increase in the quantity of information due to big data. Providing privacy is critical in this scenario. So from now on, under the knowledge of IoT's characteristics and safety demands, we need to research safety for privacy. In particular, it is essential to use bitwise operation [42] instead of mathematical algorithms such as ECC (elliptic curve cryptography) to create a lightweight safety service, encryption and authentication. Finally, it is not always necessary to protect privacy. If the customer is in an emergency condition (e.g. a vehicle accident), it is necessary to provide data on privacy to the doctor or to close individuals [43].
Multicasting security. Multicasting can be used more efficiently than standard network IoT setting that is resource constrained. Multicast group should be developed with authenticated users when using multicasting, and it is necessary to maintain secret key shared with group members [44].
Cloud. IoT systems usually use the cloud as they are unable to save the information in their small memory ability. In some instances, delicate data can be used for rescue individuals (e.g. home CCTV video, private place, health information). However, if for some reasons cloud is out of order, IoT systems are unable to save the information. Then it may be possible to miss critical information that will be used for rescue. As a consequence, the rescue service may be halted which requires the information. Accessibility is therefore extremely needed in this situation, so the unit should have back-up cloud to replace with the initial cloud.

A lot of information is sent in the cloud from many machines. Cloud should use adequate access control (e.g. authentication, permission), encryption, information anonymity, etc. to safeguard information from unauthorized users. Moreover, it may not be necessary to encrypt the information completely depending on the significance of the information. In this situation, effectiveness requires security-level based encryption. It is the comparable notion [45] with contextual integrity.

User: The user is the most vulnerable IoT security element. Even if the data system is implemented safely, any security system is ineffective if a user is careless in its treatment, particularly a system engineer. For example, when a user makes the password simple and divinable, an attacker could easily break the password by a violent force attack or a well-known security attack on the dictionary. In other words, the user must strictly follow the rules for safety and educate the user on social engineering.
Attacker: An intruder may compromise the security service even though a customer follows the safety rule. It can be a victim at any time because IoT devices are attached

to the network. Because of its limited resources, most IoT devices cannot use a powerful safety service. In addition, present IoT safety services were not fully validated. For these purposes, IoT is a simple target to attack to increase and diversify the safety attack.

In the critical control sector, the attacker will never lose control authorization and change it, and operation should be continually endorsed. Therefore, a fault tolerance [46] and back-up devices [47] must be required for the control system. When the primary unit is damaged or stopped, the backup device can be substituted.

Services: Security problems (i.e. trust, access control, middleware, storage) are analyzed in this subsection as demonstrated in Fig. 5.17. The customer must trust the server to take advantage of a service, and the server must provide the user with privacy. If the customer chooses that the server is trustworthy, the user will use smartphone, smart watch or some kind of network device service supplied by the server or group of machines. After that, bootstrapping and access control (i.e. authentication and permission) must advance with the systems. Devices get confidence from the server.

Confidence is a complicated idea. There is no clear definition, despite its significance. It is difficult to predict and assess because of the uncertain definition of trust [48]. Therefore, we need to obviously describe confidence and create the technique for estimating and assessing confidence for IoT safety.

Platform. OneM2M, OIC and other standard organizations, IoT platform requirements have been developed. Open IoT platform (for example, Mobius, OneM2M, All

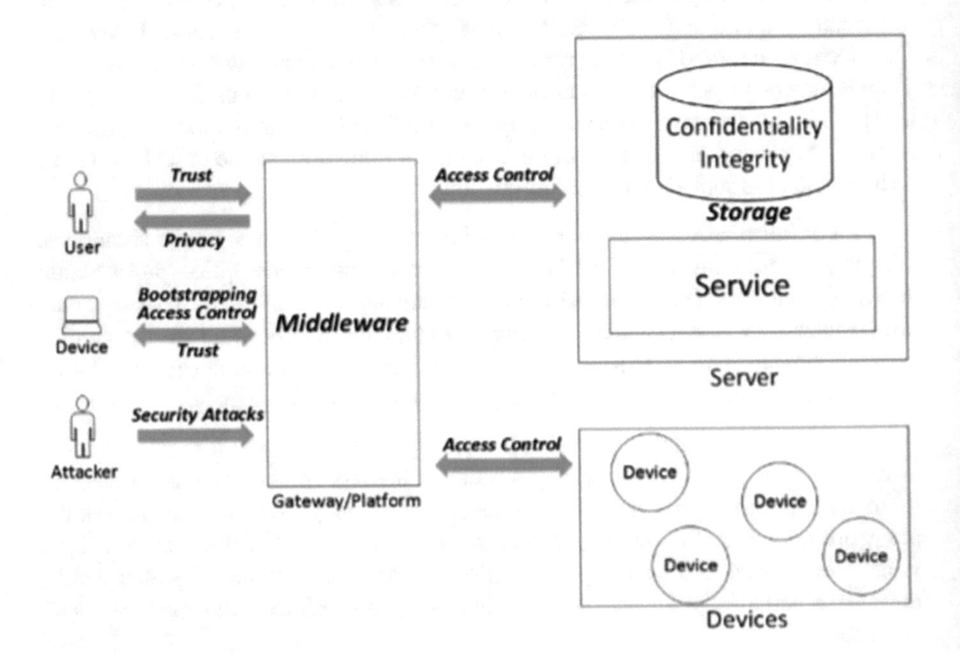

Fig. 5.17 User, device, attacker and service security requirements

Joyn, COMUS) provides multiple characteristics (for example, distributing cooperation, execution control, interoperability between heterogeneous devices with data sharing) [49]. They concentrate on platform characteristics, but security is only considered in common services (for example encryption, authentication, and signature). At this moment, as stated previously, all safety services should be lightweight because it is essential to consider the efficiency of different IoT devices. Furthermore, according to the device's safety level, security service should be optionally endorsed to resolve device performance heterogeneity. In addition, a technique for understanding many distinct safety measures used in various devices and domains is needed for authentication and permission.

5.6 Authentication in Wireless IoT

As stated above, different safety and private threats arise in an infinite stream of wireless IoT, posing severe obstacles to wireless IoT development. Many privacy-preserving authentication systems have been suggested to solve these problems and provide adequate security of privacy in wireless IoT. Identity-based authentication (IBE) enhances the procedure by altering the arbitrary public key (e.g. username). When there is only a restricted number of customers, the third party's secret can be demolished after all users receive keys. By identifying identity as a set of features (key-policy ABE) or as a set of features (cipher-text-policy ABE), attribute-based encryption (ABE) goes one step further. Only users with the relevant features can decrypt a cipher text.

Meanwhile, authentication and key agreement techniques in distinct IoT settings are also commonly studied [50]. Using a key agreement protocol, the main determination can be affected by all communicating sides and main allocation relate end issues avoided. The key-agreement methods can also enforce perfect secrecy for the future. The key agreement protocol on the basis of authentication alone is vulnerable to threats such as the MITM. In view, it was proposed to establish an authenticated key agreement for managing MITM and related attacks by a range of cryptographic authentication schemes and protocols. In this respect it was suggested to provide authenticated main agreements for dealing with MITM and associated violations with several cryptographic authenticity systems and protocols.

A gradual change to hybrid privacy-preserving authentication has taken place, combining the benefits of the processes based on symmetric-cryptosystem and asymmetric-cryptosystem. The time and spatial characteristics of the customer write process are utilized to achieve a balance of safety needs and users-friendly authentication with the initial data acquisition in the M2M communication environment [51]. Power utilization plans are aggregated from various household intelligent meters to gateway intelligent meters using holomorphic encryption and Bloom filter. Smart meters can also be used to filter messages at the Building Area Network gateway to significantly decrease the effect of traffic assault. The writers provided a hybrid cloud-based information conservation and effective data recovery system. The system adopts the common Map Reduce structure and utilizes the approach of information partitioning, independent of applications [52]. The suggested scheme's privacy-preserving function has been officially proven, and it can protect the low overhead malicious cloud model.

5.7 Recent Progress in Privacy-Conserving Authentication Wireless IoT

IoE. Multiple energy supplies are incorporated in IoE, and the interactions and complementarities of different kinds of energy present excellent difficulties in implementing mechanisms that protect privacy.

Recently, in order to coordinate various IoEs and promote their constructive interaction, study has focused on developing decentralized and scalable control and leadership logic. Ji et al. introduced a hierarchical energy management architecture and coordination based on transactive control to accomplish distribution optimization [53]. The authors focused on developing a distributed system and a decentralized economic mix to protect the privacy of each sub-area within large IoEs [54].

IoV. Unlike other IoT applications, the Internet of Vehicles (IoV) features dynamic and commonly changing network topologies as well as various communications systems, which expose IoV nodes to complicated network attacks rather than standard wireless nodes. Under different complicated communication settings, a single authentication protocol or confidence model cannot meet IoV's safety and privacy needs. IoV privacy criteria include the authentication of the car node, confidentiality of data security transferred to relevant organizations and personal routing. Specifically, in IoV, the protection of privacy in locations is an important problem in mobile apps as more precise IoV information is very important [55].

IoS. For applications such as SHS, e-health and smart cities, IoS includes multiple settings. IoS privacy protection is therefore confronted with significant problems. As the networks are growing and complex, there is a need for an appropriate privacy-control scheme with a small overhead. In recent years, much studies have been conducted concerning the design of devices which can provide not only reciprocal authentication but also traceability and the anonymity of sensor nodes. In the smart home settings involved, Kumar et al. [56] suggested an anonymous, safe structure that would provide effective authentication, allowing anonymity and unconnecting the device.

M2M Communication. The main expenses of M2M data protection schemes for wireless IoT apps are overhead, computational overhead, and storage expenses for authentication message provision. Core contracts and group authentication are suggested for speeding up the authentication process and reducing overhead signals, particularly on the resource-restricted networks. Group authentication and main contracting schemes are available to increase authentication processes and decrease overhead signaling, for resource-compressed networks.

5.8 Conclusion

The greatest challenge for WI-FI application and durability is a lack of confidence and fears among consumers regarding private leakage, robo and abuse in the IoT. The main challenge is to maintain the IoT's privacy. First, a touch brief on the definition and element of wireless IoT is present. The main IoT scenarios are subsequently provided for the applications and related models of privacy attack. Provide a short summary of the present IoT wireless authentication system for privacy pre-services. The main IoT scenarios will then be provided for applications and related models of privacy attacks. Then, we provide a short summary of the authentication of the IoT wireless devices. Although these techniques can counter diverse safety and confidentiality assaults, further attempts must be produced to secure, privileged and lower IoT wireless privacy authentication. New authentication plans were suggested and IoE, IoV, IoS and M2M were launched in the new wire authentication developments.

References

1. Rahmani, A.M., Gia, T.N., Negash, B., Anzanpour, A., Azimi, I., Jiang, M., Liljeberg, P.: Exploiting smart e-Health gateways at the edge of healthcare Internet-of-Things: A fog computing approach. Future Gener. Comput. Syst. **78**, 641–658 (2018)
2. Adeleke, J.A., Moodley, D., Rens, G., Adewumi, A.O.: Integrating statistical machine learning in a semantic sensor web for proactive monitoring and control. Sensors **17**, 807 (2017)
3. Xia, X., Chen, Z., Liu, H., Wang, H., Zeng, F.: A routing protocol for multisink wireless sensor networks in underground coalmine tunnels. Sensors **16**, 2032 (2016)
4. Caballero-Águila, R., Hermoso-Carazo, A., Linares-Pérez, J.: Optimal fusion estimation with multi-step random delays and losses in transmission. Sensors **17**, 1151 (2017)
5. Guijarro, L., Pla, V., Vidal, J.R., Naldi, M., Mahmoodi, T.: Wireless sensor network-based service provisioning by a brokering platform. Sensors **17**, 1115 (2017)
6. Wu, Z., Xu, Y., Yang, Y., Zhang, C., Zhu, X., Ji, Y.: Towards a semantic web of things: a hybrid semantic annotation, extraction, and reasoning framework for cyber-physical system. Sensors **17**, 403 (2017)
7. Choi, H.-S., Rhee, W.-S.: IoT-based user-driven service modeling environment for a smart space management system. Sensors **14**, 22039–22064 (2014)
8. Floris, A., Atzori, L.: Managing the quality of experience in the multimedia internet of things: a layered-based approach. Sensors **16**, 2057 (2016)
9. Bisdikian, C., Kaplan, L., Srivastava, M.: On the quality and value of information in sensor networks. ACM Trans. Sens. Netw. **9**, 48 (2013)
10. Suri, N., Benincasa, G., Lenzi, R., Tortonesi, M., Stefanelli, C., Sadler, L.: Exploring value of information-based approaches to support effective communications in tactical networks. IEEE Commun. Mag. **53**, 39–45 (2015)
11. Liu, J., Hu, Y., Wu, B., Wang, Y., Xie, F.: A hybrid generalized hidden markov model-based condition monitoring approach for rolling bearings. Sensors **17**, 1143 (2017)

12. Guo, Y., Chen, X., Wang, S., Sun, R., Zhao, Z.: Wind turbine diagnosis under variable speed conditions using a single sensor based on the synchrosqueezing transform method. Sensors **17**, 1149 (2017)
13. Nie, P., Dong, T., He, Y., Qu, F.: Detection of soil nitrogen using near infrared sensors based on soil pretreatment and algorithms. Sensors **17**, 1102 (2017)
14. Bellavista, P., Giannelli, C., Zamagna, R.: The PeRvasive environment sensing and sharing solution. Sustainability **9**, 585 (2017)
15. Hu, Q., Wang, S., Bie, R., Cheng, X.: Social welfare control in mobile crowdsensing using zero-determinant strategy. Sensors **17**, 1012 (2017)
16. Longo, F., Bruneo, D., Distefano, S., Merlino, G., Puliafito, A.: Stack4Things: a sensing-and-actuation-as-a-service framework for IoT and cloud integration. Ann. Telecommun. **72**, 53–70 (2017)
17. Manzalini, A., Crespi, N.: An edge operating system enabling anything-as-a-service. IEEE Commun. Mag. **54**, 62–67 (2016)
18. Amadeo, M., Molinaro, A.: CHANET: a content-centric architecture for IEEE 802.11 MANETs. In: Proceedings of the International Conference on the Network of the Future, Paris, France, 28–30 Nov 2011
19. Mendes, P.: Combining data naming and context awareness for pervasive networks. J. Netw. Comput. Appl. **50**, 114–125 (2015)
20. Delmastro, F., Arnaboldi, V., Conti, M.: People-centric computing and communications in smart cities. IEEE Commun. Mag. **54**, 122–128 (2016)
21. Chen, P.Y., Cheng, S.M., Hsu, H.Y.: Analysis of information delivery dynamics in cognitive sensor networks using epidemic models. IEEE Internet Things J. (2017)
22. Al-Turjman, F.: Energy-aware data delivery framework for safety-oriented mobile IoT. IEEE Sens. J. (2017)
23. Hamrioui, S., Lorenz, P.: Bio inspired routing algorithm and efficient communications within IoT. IEEE Netw. **31**, 74–79 (2017)
24. Pilloni, V., Atzori, L., Mallus, M.: Dynamic involvement of real-world objects in the IoT: a consensus-based cooperation approach. Sensors **17**, 484 (2017)
25. Cao, Y., Yu, W., Ren, W., Chen, G.: An overview of recent progress in the study of distributed multi-agent coordination. IEEE Trans. Ind. Inform. **9**, 427–438 (2013)
26. Caraguay, Á.L.V., Villalba, L.J.G.: Monitoring and discovery for self-organized network management in virtualized and software defined networks. Sensors **17**, 731 (2017)
27. Sotres, P., Santana, J.R., Sánchez, L., Lanza, J., Muñoz, L.: Practical lessons from the deployment and management of a smart city internet-of-things infrastructure: the SmartSantander testbed case. IEEE Access **5**, 14309–14322 (2017)
28. Uke, S.N., Mahajan, A.R., Thool, R.C.: UML modeling of physical and data link layer security attacks in WSN. Int. J. Comput. Appl. **70**(11) (May 2013)
29. Li, H., Chen, Y., He, Z.: The survey of RFID attacks and defenses. In: 8th International Conference on IEEE Wireless Communications, Networking and Mobile Computing (WiCOM) (2012)
30. Kandah, F., Singh, Y., Zhang, W., Wang, C.: Mitigating colluding injected attack using monitoring verification in mobile ad-hoc networks. In: Security and Communication Networks, pp. 1939-0122 (2013)
31. Farooq, M.U., Waseem, M., Khairi, A., Mazhar, S.: A critical analysis on the security concerns of Internet of Things (IoT). Int. J. Comput. Appl. (0975 8887) **111**(7) (Feb 2015)
32. Andrea, I., Chrysostomou, C., Hadjichristofi, G.: Internet of Things: security vulnerabilities and challenges. In: 2015 IEEE Symposium on Computers and Communication (ISCC), pp. 180–187, Larnaca (2015)
33. Chelli, K.: Security issues in wireless sensor networks: attacks and countermeasures. In: Proceedings of the World Congress on Engineering, vol. 1, 2015
34. Pathan, A.S.K., Lee, H.W., Hong, C.S.: Security in wireless sensor networks: issues and challenges. In: 2006 8th International Conference Advanced Communication Technology, vol. 2, p. 1048 (2006)

35. Modares, H., Salleh, R., Moravejosharieh, A.: Overview of security issues in wireless sensor networks. In: 2011 Third International Conference on Computational Intelligence, Modelling and Simulation, pp. 308–311 (2011)
36. Kobo, H.I., Abu-Mahfouz, A.M., Hancke, G.P.: A survey on software-defined wireless sensor networks: challenges and design requirements. IEEE Access **5**, 1872–1899 (2017)
37. Zia, T., Zomaya, A.: Security issues in wireless sensor networks. In: 2006 International Conference on Systems and Networks Communications (ICSNC06), p. 40 (2006)
38. Adedeji, Hamam, Y., Abe, B., Abu-Mahfouz, A.M.: Improving the physical layer security of wireless communication networks using spread spectrum coding and artificial noise approach. In: Southern Africa Telecommunication Networks and Applications Conference, George, Western Cape, South Africa, pp. 80–81 (Sept 2016)
39. Abdul, W., Kumar, P.: A survey on attacks, challenges and security mechanism in wireless sensor network. JIRST-International Journal for Research in Science & Technology **1**(8), 189–196 (Jan 2015)
40. Douceur, J.R.: The Sybil attacks. In: First International Workshop on Peer-to Peer Systems (IPTPS'02), (Mar 2002)
41. Zulkifli, M.Z.W.M: Attack on cryptography (2008)
42. Lee, J.-Y., Lin, W.-C., Huang, Y.-H.: A lightweight authentication protocol for internet of things. In: International Symposium on Next Generation Electronics Taiwan, pp. 1–2 (May 2014)
43. Hu, C., Zhang, J., Wen, Q.: An identity-based personal location system with protected privacy in IoT. In: EEE International Conference on Broadband Network and Multimedia Technology (IC-BNMT) China, pp. 192–195 (Oct 2011)
44. Du, X.J., et al.: An effective key management scheme for heterogeneous sensor networks. Ad Hoc Netw. **5**(1), 24–34 (2007)
45. Porambage, P., Braeken, A., Schmitt, C., Gurtov, A., Ylianttila, M., Stiller, B.: Group key establishment for enabling secure multicast communication in wireless sensor network deployed for IoT Applications. IEEE Access **3**, 1503–1511 (2015)
46. Alqassem, I., Svetinovic, D.: A taxonomy of security and privacy requirements for the internet of things (IoT). In: Industrial Engineering and Engineering Management (IEEM) Malaysia, pp. 1244–1248 (Dec 2014)
47. Roman, R., Zhou, J., Lopez, J.: On the features and challenges of security and privacy in distributed internet of things. Comput. Netw. **57**, 2266–2279 (2013)
48. Gou, Q., Yan, L., Liu, Y., Li, Y.: Construction and strategies in IoT security system. In: IEEE International Conference on Green Computing and Communications (GreenCom), 2013 IEEE and Internet of Things (iThings/CPSCom), and IEEE Cyber, Physical and Social Computing (CPSCom) China, pp. 1129–1132 (Aug 2013)
49. Sicari, S., Rizzardi, A., Grieco, L.A., Coen-Porisini, A.: Security, privacy and trust in internet of things: The road ahead. Comput. Netw. **76**, 146–164 (2015)
50. Hong, S.G., Lee, H., Choi, J.C., Bae, M.N., Lee, K.B.: Internet of things software platforms technology trends. Electron. Telecommun. Trends **30**, 39–48 (2015)
51. Zhu, H., et al.: Duth: a user-friendly dual-factor authentication for android smartphone devices. Secur. Commun. Netw. **8**(7), 1213–22 (2015)
52. Zhu, Z.G., et al.: Prometheus: privacy-aware data retrieval on hybrid cloud. In: Proceedings of IEEE INFOCOM 2013, Turin, Italy, pp. 2643–51 (2013)
53. Ji, M.J., Zhang, P.C.: Transactive control and coordination of multiple integrated energy systems. In: Proceedings of IEEE Conference on Energy Internet and Energy System Integration, Beijing, China, pp. 1–6 (2017)

54. Yang, Q.R. et al.: Decentralized security constrained economic dispatch for global energy internet and practice in Northeast Asia. In: Proceedings of IEEE Conference on Energy Internet and Energy System Integration, Beijing China, 2017, pp. 1–6
55. Zhou, X.L. et al.: ACPP: an effective privacy preserving scheme for precise location sharing in internet of vehicles. In: Proceedings of IEEE International Conference on Information and Automation, Macau SAR, China, July 2017, pp. 883–87
56. Kumar, P., et al.: Anonymous secure framework in connected smart home environments. IEEE Trans. Inf. Forensics Secur. **12**(4), 968–79 (2017)

Chapter 6
A Survey on Applications of Internet of Things in Healthcare

Naghma Khatoon, Sharmistha Roy and Prashant Pranav

Abstract Internet of Things (IoT) in healthcare is a revolution in patient's care with improved diagnosis, real examining and preventive as well as real treatments. The IoT in healthcare consists of sensors enabled smart devices that accurately gather data for further analysis and actions. By using real time data, the devices permit monitoring, tracking and management in order to enhance healthcare. Due to the efficiency of IoT, the information gathered imparts improved judgment and decreases risks of committing mistakes. IoT can also be employed for preventing machines failures which are pros as this can enhance the reliability and quality when it comes to the patient's supply chain responsiveness. This paper presents a relative survey of applications of IoT in healthcare system.

Keywords Internet of Things · Wireless Body Area Network · Wireless sensor networks · Wearable sensors · Healthcare system

6.1 Introduction

Internet of Things is a kind of wireless connectivity which connects things and people over the healthcare ecosystems to obtain, combine and examine data that is inaccessible from inside. The more commonly used cases in healthcare involve connecting people. By being able to track every patient, proper equipment and providing drugs to improve efficiency and safety by extending care beyond the four walls of the hospital and improving wellbeing by helping people make healthy choices. All of these is possible today by IoT as a result high percentage of the healthcare companies assured

N. Khatoon · S. Roy (✉) · P. Pranav
Faculty of Computing and Information Technology, Usha Martin University, Ranchi, India
e-mail: sharmistharoy11@gmail.com

N. Khatoon
e-mail: naghma.bit@gmail.com

P. Pranav
e-mail: prashantpranav19@gmail.com

V. E. Balas et al. (eds.), *Internet of Things and Big Data Applications*, Intelligent Systems Reference Library 180, https://doi.org/10.1007/978-3-030-39119-5_6

that IoT is driving the creation of new business models and the future of healthcare is nearer than we actually think. Hence, IoT is changing the healthcare system.

Now Internet is just not about connecting people i.e. Internet of People (IoP), it is about connecting things and so its name Internet of Things (IoT). The two vital keywords used are Internet and Things. Internet is a network of networks. It is a medium through which a device is connected and the thing could be any object along with intelligence to connect to the Internet like an object embedded with software or electronics. These things could include more than PC, laptop and smartphones. It could be our AC, coffee machine, cars or may be even entire city. Thus, IoT provides a junction where different devices are connected to the Internet, so that they can communicate, combine and interchange data with each other.

Further the paper is organized as follows. In Sect. 6.2, we have discussed the model of IoT for healthcare services. Section 6.3 presents the literature survey of the related works. Section 6.4 presents the system architecture and analytical model of IoT based healthcare system. In Sect. 6.5, we described the cloud based IoT healthcare systems. In Sect. 6.6, we have focused on the security and privacy concerns in IoT based healthcare system. Finally in Sect. 6.7 we have concluded the paper.

6.2 A Model for Internet of Things Healthcare Systems

The Internet of Things is the next stage of Internet. It is machine talking to machine and taking actions autonomously. The way machine sending information and taking actions accordingly, changes everything and becomes a big development especially in healthcare. Before that, healthcare system was very inefficient as information has to be moved manually, people needs to go to hospital and gather the reports. IoT now automates everything. It now makes the healthcare system automated and make it much more efficient. It saves a lot of money and so it becomes more effective also. Secondly, the important thing is the way IoT deliver care because the thing is autonomous, aware and actionable. As for example, a diabetic person may require an instant pump that would change automatically by detecting the blood sugar level and make the actions automatically. Thus, there are changes of how we instantly provide the care. The IoT enables to take many thousands of data types per second for a patient that help make decision more accurately and instantly as the information are available as well as actionable. Using IoT, as the information coming in from patients, we can now predict the disease long before they can show any sort of clinical signs and symptoms. Before IoT it can be possible in the hospital setting but outside the hospital setting with all kinds of information coming in we know more about the patient's risk factor diseases by the information of what kind of environment they live in as well as their genetic coding. Thus, by getting a lot more information, a lot more insight, a lot better practices in medicines. Some of the important applications of IoT in healthcare system are as follows:

Organs on chips: Organs on chips is one of the most pioneering inventions using IoT which facilitates making of human organs on a specific chip which in turn helps to give a thorough learning of every function of any specified organ. This function in turn increases the efficiency of any medicine. It is an advanced and a personalized treatment by analyzing what is going on in human body in real time.
Telemedicine: It is one of the most advance tool in today's life. In global healthcare, use of telemedicine becomes very important. It plays a very important role in pre-operative care and post-operative care.
3-D prescription printed bills: Another digital invention of IoT in healthcare is 3-D prescription printed bills. By employing gadgets which are wearable and other kinds devices such as sensors, a healthcare administrates print the medicine from their home by sending directions to a 3-D printer These directions are intrinsically designed as per the individual analysis of a specific for getting the proper medication doses.
Automatic wheel chairs: Automatic wheel chairs are designed using wireless networks with several sensors. This will measure and monitors several vital sign of patients and keep locations of patients at all times. This will aid in imparting safety measures to handicapped and provide them with better assistance and greater independence.
Wearable: Wearable devices in healthcare using IoT are breathtaking. Wearable nodes of a sensor measures physiological conditions. The user's pulse, blood pressure, temperature and oxygen level are the important signs for critical health evaluation. Out of the two nodes involved, the sensor node transmits data to the central node. The information thus gathered, may be implemented by employing some decisions and afterwards is forwarded to a third party. This kind of support in the sector of health care can reduce the chances of human fatigue. The detailed description of wearable healthcare system forming Wireless Body Area Network (WBAN) is described in the below sub-section.

6.2.1 Wearable Healthcare System

The greatest advantage of Internet of Things in healthcare field is to monitor, control and assist patients remotely which require the support of WBAN. Here non-obtrusive and non-invasive sensors are being focused on and sensors like implantable are excluded. The sensors are build to track and monitor basic human body parameters. Among the five basic sensors, three are used for tracking breathing/respiratory rate, body temperature and pulse rate. The other two sensors are used for measuring blood oxygen and blood pressure, which are mentioned below:

i. **Pulse sensors**

Pulse is most often considered to be very prominent criteria to discover any fatal situation. The reading of pulse can be done through various parts of a human body.

The pulse reading from a fingertip or an earlobe provides very accurate readings but may be non-wearable. On the other hand, while a chest worn system is considered to be wearable, but for a long lasting system which are wearable, wrist sensors are considered more comfortable. There are number of fitness tracking devices available with pulse measurement that include HRM-Tri by Garmin [1], H7 by Polar [2] FitBit PurePulse [3] and TomTom Spark Cardio [4].

Photoplethysmogram sensors (PPG sensors) [1] is a wearable wrist sensors used for determining pulse rate and oxygen rate in blood. The movement of pulse is checked using the sensors. The pulse rate is not recorded if the motion is high. Due to reduction of motion artifact, signal quality has been significantly improved. Authors in [5] have developed a highly-sensitive and flexible sensor for detection of pulse rate and the result shows to be promising. PPG sensors are combined with pressure sensors [6, 7] to develop pulse sensor. This sensor reads pulse rate from multiple points at wrist for giving clear and accurate values that helps in successful diagnostics. Hence, PPG sensors proves to be effective device for reading pulse rate accurately with reduce noise on the signal quality.

ii. **Breathing/Respiratory rate sensors**

Based on the previous research works and their level of significance in designing sensors for measuring breathing/respiratory rate, the first one is thermistor, which is considered as nasal sensors. However, Echocardiogram (ECG) signals is also used for measuring the same, popularly known as ECG Derived Respiration (EDR) [8]. Purpose of this signal is to estimate breathing patterns and find apnea events. This method gets the study of rate of respiration very well and is wearable. The contact of ECG devices sometimes may cause irritation to the skin on long use and as they are non-reusable so they need to be changed periodically. In [9], microphone has been used to detect respiration and for measuring respiratory rate, which focuses on detecting wheezing which is a very common symptom of asthma. However, due to external noise, microphone is not a good choice for wearable device.

iii. **Body temperature sensors**

One of the useful human body parameters is body temperature and can be considered as wearable healthcare system as it helps in diagnosis of hypothermia, heart stroke, fevers and many more. Nowadays, thermistor-type sensors are deployed for measuring body temperature. In [10, 11] the body temperature was measured using Negative-Temperature-Coefficient (NTC) type sensor, whereas, in [12, 13] for measuring body temperature Positive-Temperature-Coefficient (PTC) type of sensors were taken into consideration. However, accuracy of determining the body temperature is restricted to the accurate placement of sensors in human body.

iv. **Blood pressure**

Blood Pressure (BP) is an often measured vital sign along with others. Hypertension is one of the major risk factor for cardiovascular diseases consisting heart attack. So, amalgamation of BP with the WBAN for healthcare purpose will impart important information for patients. Yet there is a challenge of designing a wearable sensor

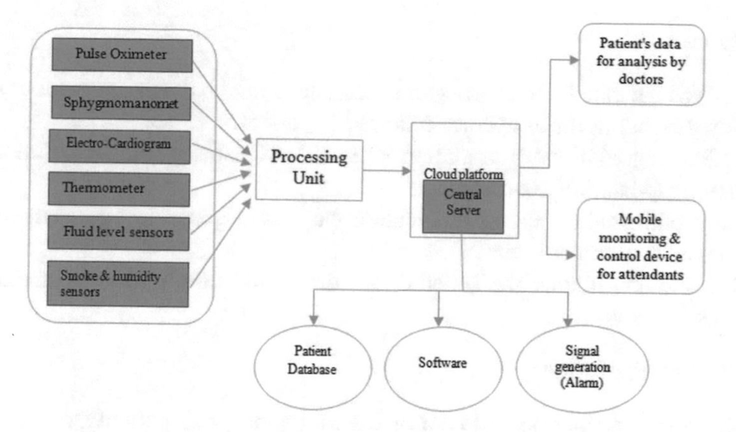

Fig. 6.1 Block diagram of IoT based healthcare system

which can be used for continuously and non-invasively monitoring BP. Further study can be done to develop a device which can easily and comfortably measure blood pressure as yet no device has been developed to accurately measure BP.

v. **Pulse oximetry sensors**

Oxygen level in blood is measured using pulse oximetry sensors. The level of blood oxygen is determined by generating PPG signals. Generally there are two LEDs—one red and the other one is infrared, both are administer through the skin. Initially, pulse oximeters were wore on the finger clip but now several studies are done to develop devices which are more portable. The pulse oximeter that was designed was having low power in order to improve the wearability. Under extreme situation when a patient undergoes shock, hypothermia or other conditions, an in-ear reflexive pulse oximeter can detect oxygen level in blood. Figure 6.1 shows an IoT based healthcare system which is summarized as below.

- Product Infrastructure:
  - Hardware (Amplifiers for sensor signals + Dedicated device for displaying data)
  - Software (Control algorithms).
- Sensors
  - Pulse Oximeter
  - Sphygmomanometer (Blood Pressure)
  - Electro-Cardiogram
  - Thermometer
  - Fluid Level Sensor
  - Humidity and Smoke Sensor.
- Connectivity
  - Bluetooth (From sensor to Microcontroller)
  - Wi-Fi (Microcontroller to Server and Server to Device).

- Analytics
  - Analyze the data from sensors and correlate to get healthy parameter of patient.
  - Alert Attendant for assistance required to patient.
  - Predict degradation/up-gradation of health of patient using continuous data learning (Machine Learning).
  - Study and predict wrong medication cases and generate an alarm for body changes due to medicines.
  - Generate and update the schedule for doses and timely alert to attendant for doses.
- Application Platform
  - Access of information to doctor on his computer for all patients with all details.
  - Dedicated monitoring device for attendants.

Some other applications of IoT in healthcare include:

- Sensor enabled pills—these pills can provide information about the drug intake.
- Monitoring of Cardiac activity.
- Medication dispensation e.g. smart, connected insulin pumps.
- Data collection from human body at real time such as fitness trackers, sleep trackers etc.
- Diagnostics and monitoring of patients remotely.
- Clinical care—collection of physiological information of patients during critical condition.
- Living space monitoring
- Monitoring of stock level for medical supplies.

6.2.2 IoT Benefits in Healthcare Services

Some of the advantages of IoT in healthcare is depicted in Fig. 6.2. The first and foremost benefit of IoT healthcare system is to reduce the healthcare cost. Presently, the healthcare cost is one of the major issue of concern for everyone. However, it's not only the little healthcare cost involvement, the worst situation of today's is that the healthcare is reactive in nature. The reactive healthcare means that the people are going to react on any disease when the symptoms arises i.e. the reactive healthcare involves reacting to an adverse situation or condition of any disease. However, we should invest more on the preventive healthcare i.e. the proactive healthcare. The proactive healthcare differs from reactive healthcare in the sense that actions in the former are taken before the symptoms get marked instead of waiting till the symptoms really aroused as seen in the later one. Thus, if we keep on spending on the reactive healthcare, the healthcare cost continue to grow. With the IoT, the healthcare cost will be reduced. This is because with IoT the data will be available immediately to

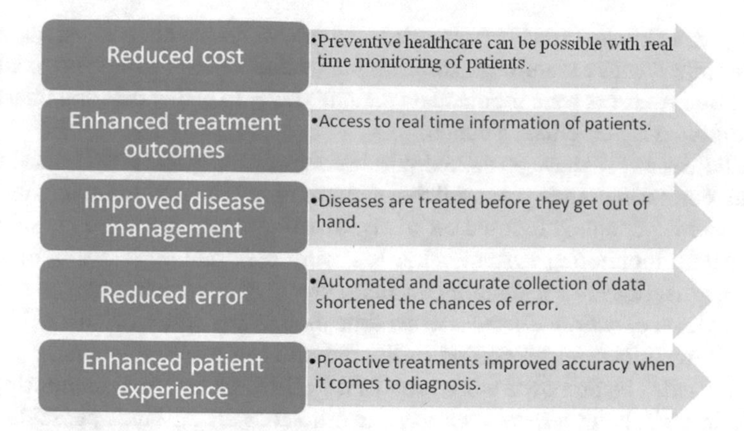

Fig. 6.2 Importance of IoT healthcare services

the doctor or the people and appropriate actions can be taken. Secondly, with IoT, beautiful outcome of every treatment become possible because the doctor now get real time information and access to everything. The IoT will reduce the healthcare cost, management cost and better monitor the patient's health condition. The IoT in healthcare also reduce the disease management. With IoT the chances of medical error reduced.

Recent research in medical reveals that medical errors are the third leading cause of death that is even more than any aviation death or road traffic accidents. Incidents like a patient getting a wrong medicine, a surgeon operating a wrong body part, overdoses of medicines and other medical mistakes are more common. The scientific study of patient's safety recognize that people don't just die from bacteria and cancer. They also die from wrong medical treatment. With the IoT, the access to real time information of patients reduced such medical errors.

The IoT enhance the patient delivery also because with IoT the information is available everywhere and they can easily verify and track the things on their television, refrigerator or any other connected device.

6.3 Review of Literature

Some of the literature surveys of IoT in healthcare applications are listed in this section. The health monitoring using IoT is one of the recent research area. Niranjana [14] has given an intelligent home-based healthcare IoT system where he employed a Medical Box (iMedBox) and gateway called iGATE. The iMedBox and iGATE were used for home healthcare IoT system and gateway respectively. Wearable sensors are integrated with intelligent medicine packaging (iMedPack) by using wireless networking system. Radio Frequency Identification (RFID) was used to couple iMedPack with iMedbox benefiting users to get real time healthcare prescriptions. Alharbe et al. [15] also presented application of ZigBee and RFID technologies in healthcare in conjunction with the internet of things.

Yvette and Gelogo [16, 17] have presented the IoT U-healthcare system. In their work, mobile devices were used for communication of data to doctor or medical stations. Body sensors were mounted on patient's body to gather data and transmitted to be examined at surveillance center.

Castillejo et al. [18] have presented Wireless Sensor Network (WSN) based IoT for E-health monitoring. They proposed the incorporation of wearable devices on WSN. Their works were mainly focused on arising alarming alerts in hazardous situations. This E-health monitoring system using IoT were designed by focusing on certain specific applications like generating an alarm signal as and when a person reaches a hazardous level of action. This was a real time monitoring system of one's life which stop them to further get proceeded in the direction of exercise when an alarming signal aroused. The proposed work provide a well-tested real time monitoring in E-health using IoT, however, the accuracy and efficiency are still indispensable causes of concern.

He et al. in their paper [19] have presented the concern of security for WBAN using emerging technologies with IoT. The efficiency of healthcare using IoT with WBAN improved the modern medical system as the real time collection of data from different sensors incorporated using WBAN directly from patient's body anytime anywhere. They focused on several security and confidentiality issues of IoT based healthcare monitoring. Different schemes in this regard have been proposed, however, security is still a major concern and a challenging issue.

Zhang et al. proposed the way to improve the lifetime of WBAN so as increase the duration of real time data gathering by sensor networks in IoT based healthcare system [20]. In IoT based healthcare using WBAN the sensors play a major role of collecting and transmitting the real time data. Their energy depletion makes the healthcare data collection less reliable. As the sensors are battery dependent and if those devices are not managed properly the energy depletion degrade the overall network lifetime. The authors in their work have proposed power game-based approach for efficacious management of power control in WBAN and thus improving the social interaction.

Arcadius et al. surveyed the IoT based healthcare applications and challenges [21]. As the applications of IoT is growing and in future it will be a pre-requisite along with several digital technologies, focus is given on introducing new Internet Protocol addressing which can satisfy any node in the network.

Dhanvijay et al. in their paper [22] have done survey of enabling technologies in healthcare and its applications. The authors scrutinize various healthcare applications and WBAN based IoT healthcare technologies. The IoT healthcare network devices and sensors have limited battery power, and so the requirement of power efficient sensors for e-healthcare using IoT is focused upon.

Authors in [23] tried to portray the objectives of IoT and how it has brought a revolution to the concept of Internet by connecting not only human but also M2M connection. This paper addressed the basic IoT architecture focusing its importance as compared to regular information network. Different challenges arising out from technical, economical and application aspects are also discussed in the paper. Despite of its several challenges IoT has find its application in day to day life starting from smart home, smart city, smart healthcare systems, supply chain management, smart agriculture etc. Supporting the benefits of IoT the authors also presented case studies on Multi Robot system where interaction and data transmission is done between robots for executing mission against terrorism. Another application of IoT is seen in hospitals emergency section where RFID tags are used for providing urgent medical treatments to critical patients. The paper is concluded by highlighting the three visions of IoT and those are things oriented vision, Internet oriented vision and semantic oriented vision.

Authors in [24] have presented different technological evolutions of IoT, mentioning the various research trends of IoT. The paper also presented the service-oriented architecture of an IoT mentioning different layers and their functionalities. The authors have identified different emerging technologies where IoT can be implemented including several benefits and challenging issues. It has been observed that applications of IoT is found in diversified fields starting from healthcare, social network, business and industry applications.

Looking into the demand of low cost smart hospitals authors in [25] have proposed a new technological invention which is cheaper and beneficial for development of smart hospitals. The proposed framework helps in patient monitoring remotely and also helps in indication of alerts during patient's critical health condition. The use of sensors in wearable devices helps in monitoring patient's health condition from remote location. Also sensors are attached to patient beds which help in recording patient's each and every vital sign for better medical treatment and analysis. The use of IoT helps in reducing the cost of the proposed architecture. Huge amount of data is collected which is stored in cloud and analyzed using big data technique. Thus the proposed system is effective for better medical treatment with low cost and improved quality.

In healthcare applications Wireless Body Area Network (WBAN) is attaining much more importance because of implantation of wearable devices in human body. Authors in [26] have proposed an autonomous WBAN having three features that includes a solar panel for solar energy harvesting with MPPT circuit, Bluetooth energy transmission with wearable sensor node and a smart phone for visualizing data and giving notification in case of emergency. Different wearable sensors can be placed in human body for measuring the body temperature, heartbeat and other parameters which can be notified using smart phone. The use of solar energy harvester acts as a power supply which will extend the lifetime of the sensor nodes. Moreover, the wearable devices are also built in such a way keeping in view the usability and ease-of-use feature from user's perspective.

Authors in [27] have conducted a qualitative study for adoption of WSN-based smart home healthcare systems (WSN-SHHS) which addresses facilitators and barriers by indicating cognitive, socio-technical, socio-economical factors. Literature survey of WSN-SHHS focused on its technological aspects giving less observation on the patient's issues for adoption and use. Thus, authors in this paper presented a mixed method for providing great insights on patient's perception about WSN-SHSS system using both quantitative and qualitative research methods. Quantitative method uses survey technique based on certain questionnaires prepared from hypothesis designed by the authors. Whereas, qualitative method uses Interview technique. The result analysis focuses on certain findings and guidance which need to be followed for the design and development of WSN-SHSS. However, the research study also finds out certain limitations and its scope for future work.

Authors in [28] have presented a model for implementation of WBAN in a healthcare system highlighting the design features that need to be incorporate in WBAN system. The proposed model targets multi-patient monitoring in medical centers focusing on individual use, use in one room and one floor. The paper described the hardware implementation of the model using sensor nodes and CCU. The data collected from the model is transmitted to some remote locations using multi hoping technique which uses MAC protocols. This system helps in monitoring patient's physiological parameters remotely. The proposed system is different from the traditional system in the sense that it can monitor many patients' physiological signals to have a feeling of real hospital environment.

Authors in [29] have proposed a cooperative IoT model for monitoring and controlling health of rural people. The system provides effective Quality-of-Service (QoS) because of the cooperation technique. The result analysis shows the trade-offs of energy, latency and throughput using NS-2 simulation tool. The proposed model prove to be reliable for continuous health monitoring. In future, the author proposed to expand the model by adding authentication and authorization feature in the cooperative IoT system. Pantelopoulos et al. [30] presented a survey on wearable sensor-based systems for health monitoring and prognosis where Wearable Health Monitoring System (WHMS) have potential to revolutionize by providing low cost, early detection and better treatment of chronic diseases. Pervez et al. [31] presented a comprehensive discussion on the medical applications of wireless body area networks in smart healthcare systems including epileptic seizure warning, glucose monitoring and cancer detection and proved that WSN and WBAN in medicine can further expanded to improve the quality of life. Huang et al. [32] have presented a healthcare monitoring architecture coupled with wearable sensor systems and an environmental sensor network for monitoring elderly or chronic patients in their residence. Donovan et al. [33] have designed a body area network for falls assessment among elder patients with an emphasis on the communication scheme chosen. Boavida et al. [34] in their paper have presented a review on the challenges, approach and enabling technologies on people centric IoT. In [35], the authors have proposed a new seeding technique that uses data collected from sensors that can be used as a seed for Random Number Generators (RNG). This technique can increase the security of smartly connected devices without any significant increase in the cost incurred to deploy such a generator. Authors in [36] have introduced a new framework for the evaluation purpose of the Health Information System (HIS) that is fit technically as well as from human and organizational perspective. The result shows that some positive employment of human attitude and skill base together with good leadership will positively impact the proper adoption of HIS.

6.4 System Architecture and Analytical Model of IoT Based Healthcare System

After going through the broad variety of IoT based healthcare systems, needs for the designing of such systems become apparent. In the healthcare systems using IoT as their architecture, the data gathered from sensors, prescriptions and from previous medical data will be used for taking descriptive, diagnostic and predictive and event based actions as shown in Fig. 6.3. Each paper focuses on using sensors for keeping the record of patient's current health condition. The authors in [37, 38] suggest the deployment of home sensors based on characteristics gathered from environment vision. This in turn restricts to a specific location the use of the IoT system. So, the suggestion of deploying all the sensors to be portable with wearable nodes which are fixed externally was done. This will enable to monitor patient's health wherever they go by imparting non-intrusive, safe and restful solutions and the patients would therefore turn out to be more responsive by using the technology which is continuously monitored based on the health. Repairing and replacing nodes which are externally wearable would be much simpler as compared to the sensors fixed at one specific location. The use of Bluetooth which is a short range communication system is preferred in the current state of the art for sending data from sensors to a Smartphone which requires further processing. LTE, a long range communication system can be deployed to transfer the gathered processed information via SMS or Internet to the healthcare provider from the patient. The restriction is that smartphones have

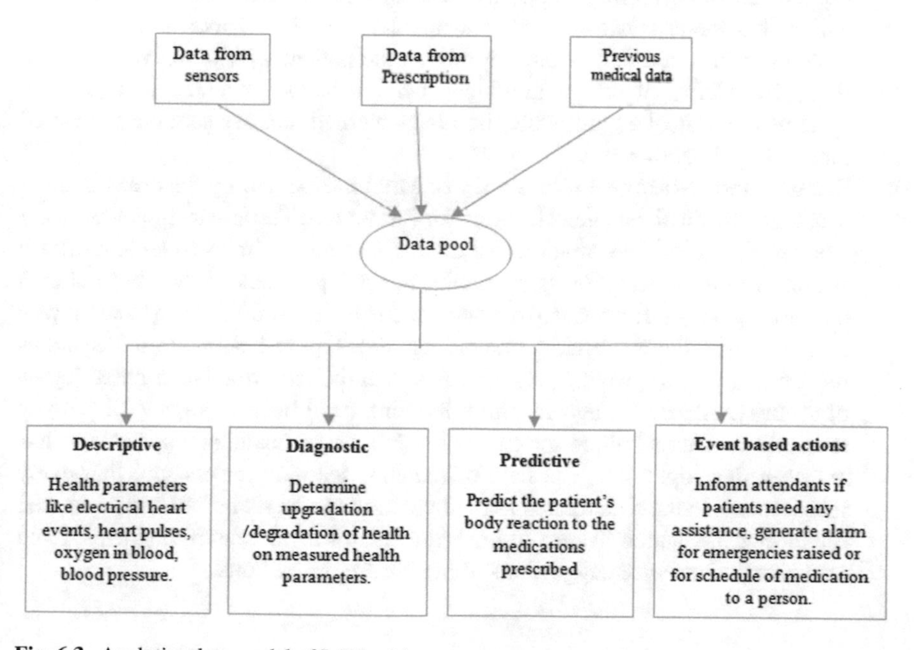

Fig. 6.3 Analytics data model of IoT healthcare system

limited battery life and require periodic recharging. So, a patient with a discharged battery would be disconnected from all the healthcare centers. In such scenario, the use of low-powered node is recommended. Such nodes such be specially designed for gathering and managing information from the healthcare system. Storing high volumes of medical data requires cloud storage which is also mentioned in research works related from healthcare system [39, 40]. Based on the repeating trends in the literature till date the following will support the development of future Internet of Things in the healthcare systems:

i. **Short-Range Communications**: A short-range communication technology is needed for sensors to interact with the central node. Apart from human body, security and latency, many such things need attention for choosing a standard for communication in the shorter range. Further, the deployment of any such method must not impact on the human. The method also must be incorporated with strong security mechanism so that the private data of the patient must be protected. At last low-latency is significant for time-critical systems such as a system that tracks critical health and also calls an ambulance if the need arises.
ii. **Long-Range Communications**: Collection of raw data by the central node is useless unless something useful is generated from the data. Rather, these data must be forwarded to a central hub from where these can be processed by a caretaker/doctor. As in short range systems, there are many constraints which must be given a thorough study for choosing a communication standard for the long range system deployed in healthcare. Some factors include security, capabilities of correcting errors; dynamicity against interference, low-latency, and such systems must be readily available. The readily availability of such systems ensure that that information from the patient will be delivered all the time, irrespective of the geographical location of the patient. Although such systems are critical as far as the time is concerned, but are suitable for varied range of applications.
iii. **Secure Cloud Storage Architecture & Machine Learning**: Secure storage of a patients' medical data must be done for further use. The advantage of knowing a patients' medical information to a doctor is of no use unless a large database storing all the information is available. Several previous studies have shown that storing such information over the cloud is the most reliable and secure way available. In today's scenario, researchers developing healthcare IoT systems must find a way to provide professionals with the required data a great degree of security. Also, although machine learning have been recognized by many researchers as a technique for enhancing IoT based healthcare systems, it has not been developed yet. The field of machine learning makes sure that many recent trends in the data related to medical field can be identified and processed for further use which in turn offers better plan for treatment, diagnostics and also give patient-wise suggestions to healthcare professions.

6.5 Cloud Based IoT Healthcare Systems

The use of cloud storage in the processing and analysis of big data makes cloud computing a much researched field. Many previous works have been technically reviewed many times to design systems supporting healthcare which are based on cloud computing framework.

Some of the benefits provided by cloud technologies in the healthcare are listed below:

- Software as a Service (SaaS)—with the use of SaaS, the providers of healthcare functionality may practice with the data of relevance and perform other useful tasks.
- Platform as a Service (PaaS)—this enables users with many tools and techniques the purpose of network management, database management and virtualization.
- Infrastructure as a Service (IaaS)—it gives the infrastructure which is physical in nature for storage and server purposes.

These services can be practiced to attain variety of tasks but two most significant uses can easily be identified.

i. **Big Data Management**

The 5 V's of a big data management are volume, velocity, variety, veracity and value. The amount of data which are obtained and then after generated is given by volume while velocity gives the speed at which these data are generated. Variety is the variance in the type of data while veracity is the uncertainty that later what kind of data will be added and value is the information that can be gained from the big data set. Main drawback of big data management lays in the fact that it cannot design a system with specific features to a big data set. Considering the volume and variety of data needed for developing a emergency healthcare IoT based system, a data storage model can be designed with the main objective of arranging data of different types easily accessible of Its objective is to arrange heterogeneous physiological data and make them easily accessible to relevant healthcare providers during emergency situations. The benefits of cloud technologies for big data management are as follows:

- The storage space provided by big data is tremendous and can be said to be almost unlimited.
- It provides the facility of many highlighting services and is readily accessible by both the patient and the doctor. This gives the patient a continuous monitoring of their own healthcare.
- It facilitates the doctors of providing a more accurate and suitable treatment without meeting in person with the patient.

ii. **Data Processing and Analytics**

Computational offloading and machine learning are the two data processing technologies out of many such technologies provided by a cloud based IoT system. Computational offloading necessarily employs cloud to execute such data which are beyond the resources of wearable devices with limited resource capacity. On the other hand, through machine learning, the collection of information from a large data set is quite easy. ML can associate meaningful links between symptoms and diseases, and hence can predict accurately some possible diagnoses based on those provided to previous patients, developing patient-wise treatment plan on the basis of determining such plans which has worked for patients with similar illness in the past and much more.

6.6 Security and Privacy Issues in IoT Healthcare

Anything that gets connected inherits the same threats as our computers or our smartphones. So we need to place the same concern for IoT security. Security is being identified as top concern for doctors. Some IoT healthcare security issues are discussed below:

- Ransomeware can target devices which are connected: end point security will be more important than ever. Healthcare organization would be main target by Ransomeware because they rely on private data records to deliverable healthcare systems. Such data are crucial and are shared by patients to their respective IoT based healthcare and hence devices making such devices a prime target for attack.
- Threats can be minimized by educating the insiders: employees cause majority of threats and malicious activities in the healthcare system. Without the use of very accurate analysis tool it is actually impossible to differentiate between well intentional employees and the one who are outsiders.

However, privacy and security are two indispensable issues in healthcare industry. Healthcare is more sensitive to privacy aspects specially. It's a bigger barrier to overcome. The other things are risk because we can have automatic or autonomous machines that can create harm to someone and the last one is value as we should try to decrease cost without compromising with quality.

The IoT includes multiple of domains and unlimited of devices which are getting connected on a daily basis. The healthcare and the general practice of medicine majorly face issues in one or more of the following three things:

- Non-availability of real time data: Medical research has to rely mostly upon the data which are residual for medical examinations. Raw data which can solve many complex medical conditions is not available in the healthcare. These issues can be tackled with the use of IoT. Through IoT, one can get access to tremendous of data through a proper analysis and testing in the real time of these data can be done.
- Lack of smart care devices: IoT empowers healthcare professionals and improves the quality of care. Finally, it brings down the high cost of medical devices. For example the working of a smart care device which consists of certain parameters that are considered safe. Once any one of the parameter is breached, the sensors immediately relay this message via a secure gateway to a cloud. The cloud then

passes a remote signal to a smart device that is monitored either by a nurse or a caretaker at home. The beauty of remote patient monitoring is that patients can now replace a long wait at the doctor's office with the quick checking, data shares and instructions on how to proceed. IoT hence, bridges the gap between reading devices and delivering healthcare by creating systems rather than just equipment.
- Cloud security and privacy: In cloud-based systems, security is a vital issue. In the healthcare systems based on cloud, it is essential that the information related to the health of the patient can be accessed easily by all the relevant parties at the same time and at the same time it is equally important to keep the patient's sensitive health data private. Few researches have been done for developing security protocols dynamic enough for healthcare practices. An access control scheme named "Safe-Protect" is proposed, with the main motive of giving the patient the full control of their private data. The individual patient creates a policy that allows specific healthcare providers to access and get their health record. The individual patient's data is encrypted and stored in cloud storage. If any healthcare provider wants to access the patient's health record, they must have their credentials.

6.7 Conclusion

The future of healthcare system lies with the Internet of Things (IoT) especially WBAN. With enhanced data regarding patients and real time tracking, the hopes of preventing and curing diseases are higher than ever. Such a small but an extremely smart piece of digital technology is rapidly changing the traditional approach of healthcare and treatments. Production of healthcare product is easing the patient's trouble along the way. This paper presents the applications of IoT in healthcare services and examines the different issues of research activities in this regard. Real time collection and examination of data helps in the early detection and prevention of diseases. However, instead of the resourceful applications of IoT in healthcare there are certain areas that needs more research consideration like security, privacy and power resource management in order to make the healthcare system more robust, reliable and scalable. With the emergence of various technologies of Wireless Body Area Network (WBAN) and IoT devices enables the monitoring, medication and support of home health services more effective. Hence, IoT has a significant and profound future research work in healthcare.

References

1. Garmin: HSM-Tri. Available at: https://buy.garmin.com/en-AU/AU/p/136403 (2017)
2. Polar: H7 Heart Rate Sensor. Available at: https://www.polar.com/auen/products/accessories/H7heartratesensor (2017)
3. FitBit: FitBit PurePulse. Available at: http://www.fitbit.com/au/purepulse (2017)

4. TomTom: TomTom Spark Cardio. Available at: https://www.tomtom.com/en_au/sports/fitness-watches/gps-watch-cardio-spark/blacklarge/ (2017)
5. Shu, Y., Li, C., Wang, Z., Mi, W., Li, Y., Ren, T.L.: A pressure sensing system for heart rate monitoring with polymer-based pressure sensors and an anti-interference post processing circuit. Sensors (Basel, Switzerland) **15**(2), 3224–3235 (2015)
6. Wang, D., Zhang, D., Lu, G.: A novel multichannel wrist pulse system with different sensor arrays. IEEE Trans. Instrum. Meas. **64**(7), 2020–2034 (2015)
7. Wang, D., Zhang, D., Lu, G.: An optimal pulse system design by multichannel sensors fusion. IEEE J. Biomed. Health Inform. **20**(2), 450–459 (2016)
8. Varon, C., Caicedo, A., Testelmans, D., Buyse, B., Huffel, S.V.: A novel algorithm for the automatic detection of sleep apnea from single-lead ECG. IEEE Trans. Biomed. Eng. **62**(9), 2269–2278 (2015)
9. Oletic, D., Bilas, V.: Energy-efficient respiratory sounds sensing for personal mobile asthma monitoring. IEEE Sens. J. **16**(23), 8295–8303 (2016)
10. Aqueveque, P., Gutierrez, C., Saavedra, F., Pino, E.J., Morales, A., Wiechmann, E.: Monitoring physiological variables of mining workers at high altitude. IEEE Trans. Ind. Appl. **PP**(99), 1 (2017)
11. Narczyk, P., Siwiec, K., Pleskacz, W.A.: Precision human body temperature measurement based on thermistor sensor. In: 2016 IEEE 19th International Symposium on Design and Diagnostics of Electronic Circuits & Systems (DDECS), pp. 1–5 (2016)
12. Nakamura, T., Yokota, T., Terakawa, Y., Reeder, J., Voit, W., Someya, T., Sekino, M.: Development of flexible and wide-range polymer-based temperature sensor for human bodies. In: 2016 IEEE-EMBS International Conference on Biomedical and Health Informatics (BHI), pp. 485–488 (2016)
13. Eshkeiti, A., Joyce, M., Narakathu, B.B., Emamian, S., Avuthu, S.G.R., Joyce, M., Atashbar, M.Z.: A novel self-supported printed flexible strain sensor for monitoring body movement and temperature. In: IEEE SENSORS 2014 Proceedings, pp. 1615–1618 (2014)
14. Niranjana, S., Balamurugan, A.: Intelligent E-health gateway based ubiquitous healthcare systems in Internet of Things. Int. J. Sci. Eng. Appl. Sci. **I**(9) (2015). ISSN: 2395-3470
15. Alharbe, N., Atkins, A.S., Akbari, A.S.: Application of ZigBee and RFID technologies in healthcare in conjunction with the internet of things. In: Proceedings of the International Conference on Advances in Mobile Computing & Multimedia, p. 191 (2013)
16. Yvette: Internet of Things (IoT) for U-healthcare. Adv. Sci. Technol. Lett. **120**, 717–720 (2015)
17. Gelogo, Y.E., Hwang, H.J., Kimz, H.-K.: Internet of Things (loT) framework for u-healthcare system. Int. J. Smart Home **9**(11), 323–330 (2015)
18. Castillejo, P., Martinez, J.F., Rodriguez-Molina, J., Cuerva, A.: Integration of wearable devices in a wireless sensor network for an E-health application. IEEE Wirel. Commun. **20**(4), 38–49 (2013)
19. He, D., Zeadally, S., Kumar, N., Lee, J.H.: Anonymous authentication for wireless body area networks with provable security. IEEE Syst. J. **11**(4), 2590–2601 (2016)
20. Zhang, Z., Wang, H., Wang, C., Fang, H.: Interference mitigation for cyber-physical wireless body area network system using social networks. IEEE Trans. Emerg. Top. Comput. **1**(1), 121–132 (2013)
21. Arcadius, T.C., Gao, B., Tian, G., Yan, Y.: Structural health monitoring framework based on internet of things: a survey. IEEE Internet Things J. **4**(3), 619–635 (2017)
22. Dhanvijay, M.M., Patil, S.C.: Internet of Things: a survey of enabling technologies in healthcare and its applications. Comput. Netw. **153**, 113–131 (2019)
23. Perumal, K., Manohar, M.: A survey on internet of things: case studies, applications, and future directions. In: Internet of Things: Novel Advances and Envisioned Applications, pp. 281–297. Springer (2017)
24. Li, S., Xu, L.D., Zhao, S.: The internet of things: a survey. Inf. Syst. Front. **17**(2), 243–259 (2015)
25. Rizwan, P., Suresh, K.: Design and development of low investment smart hospital using Internet of things through innovative approaches. Biomed. Res. **28**(11), 4979–4985 (2017)

26. Wu, T., Wu, F., Redoute, J.M., Yuce, M.R.: An autonomous wireless body area network implementation towards IoT connected healthcare applications. IEEE Access **5**, 11413–11422 (2017)
27. Alaiad, A., Zhou, L.: Patients adoption of WSN-based smart home healthcare systems: an integrated model of facilitators and barriers. IEEE Trans. Prof. Commun. **60**(1), 4–23 (2017)
28. Yuce, M.R.: Implementation of wireless body area networks for healthcare systems. Sens. Actuators A **162**(1), 116–129 (2010)
29. Rohokale, V.M., Prasad, N.R., Prasad, R.: A cooperative internet of things (IoT) for rural healthcare monitoring and control. In: Proceedings of the 2nd International Conference on IEEE Wireless Communication, Vehicular Technology, Information Theory and Aerospace & Electronics Systems Technology, pp. 1–6 (2011)
30. Pantelopoulos, A., Bourbakis, N.G.: A survey on wearable sensor-based systems for health monitoring and prognosis. IEEE Trans. Syst. Man Cybern. Part C Appl. Rev. **40**(1), 1–2 (2010)
31. Pervez Khan, M., Hussain, A., Kwak, K.S.: Medical applications of wireless body area networks. Int. J. Digit. Content Technol. Appl. **3**(3), 185–193 (2009)
32. Huang, Y.M., Hsieh, M.Y., Chao, H.C., Hung, S.H., Park, J.H.: Pervasive, secure access to a hierarchical sensor-based healthcare monitoring architecture in wireless heterogeneous networks. IEEE J. Sel. Areas Commun. **27**(4), 400–411 (2009)
33. O'Donovan, T., O'Donoghue, J., Sreenan, C., Sammon, D., O'Reilly, P., O'Connor, K.A.: A context aware wireless body area network. In: Proceedings of the 3rd International Conference on IEEE Pervasive Computing Technologies for Healthcare, Pervasive Health 2009, London, UK, pp. 1–8 (2009)
34. Boavida, F., Silva, J.S.: IoP—internet of people. In: Proceedings of the Future Internet Networking Session, ICT2013, Vilnius, Lithuania (2013)
35. Hong, S.L., Liu, C.: Sensor-based random number generator seeding. IEEE Access **3**, 562–568 (2015)
36. Yusof, M.M., Kuljis, J., Papazafeiropoulou, A., Stergioulas, L.K.: An evaluation framework for health information systems: human, organization and technology—fit factors. Int. J. Med. Inf. **77**(6), 386–398 (2008)
37. Zhu, N., Diethe, T., Camplani, M., Tao, L., Burrows, A., Twomey, N., Kaleshi, D., Mirmehdi, M., Flach, P., Craddock, I.: Bridging e-Health and the Internet of Things: the SPHERE Project. IEEE Intell. Syst. **30**(4), 39–46 (2015)
38. Pasluosta, C.F., Gassner, H., Winkler, J., Klucken, J., Eskofier, B.M.: An emerging era in the management of Parkinson's disease: wearable technologies and the internet of things. IEEE J. Biomed. Health Inform. **19**(6), 1873–1881 (2015)
39. Dimitrov, D.V.: Medical internet of things and big data in healthcare. Healthcare Inform. Res. **22**(3), 156–163 (2016)
40. Zenko, J., Kos, M., Kramberger, I.: Pulse rate variability and blood oxidation content identification using miniature wearable wrist device. In: 2016 International Conference on Systems, Signals and Image Processing (IWSSIP), pp. 1–4 (2016)

Dr. Naghma Khatoon is currently working as Assistant Professor, Dept. of Computer Science at Usha Martin University, Ranchi. She received her B.Sc. IT from Ranchi University and M.Sc. IT and Ph.D. from Birla Institute of Technology, Mesra in the year 2010, 2012 and 2018 respectively. Her research interest is in the area of WSN, MANET, IoT.

Dr. Sharmistha Roy is working as a Assistant Professor, in Faculty of Computing and Information Technology, Usha Martin University, Ranchi, India. Moreover, she has received Gold medal during her M.Tech. Her interests focus on Cloud Usability, Security, and Software Engineering.

Mr. Prashant Pranav is currently pursuing Ph.D. from BIT Mesra. His expertises include Cryptography, Algorithm Analysis, and Computational Musicology. He has completed M.E (Software Engineering) from BIT Mesra and B.Tech in Computer Science and Engineering from College of Engineering Bhubaneswar.

Chapter 7
Mobile Applications Dedicated for Cardiac Patients: Research of Available Resources

Gonçalo F. Valentim Pereira, Ivan Miguel Pires, Gonçalo Marques, Nuno M. Garcia, Eftim Zdravevski, Petre Lameski, Francisco Flórez-Revuelta and Susanna Spinsante

Abstract In recent years cardiac problems and using mobile devices for aiding people with these problems have received significant attention from the scientific communities to develop solutions to improve the quality of life. The proliferation of mobile computing technologies has revolutionized the medical practices in both patient and clinical staff sides. In particular, the development of mobile health applications continues to increase; mainly, the cardiology field is the most addressed.

G. F. Valentim Pereira · I. M. Pires · G. Marques (✉) · N. M. Garcia
Computer Science Department, Universidade da Beira Interior, Covilhã, Portugal
e-mail: goncalosantosmarques@gmail.com

G. F. Valentim Pereira
e-mail: goncalo.pereira@ubi.pt

I. M. Pires
e-mail: impires@it.ubi.pt

N. M. Garcia
e-mail: garcia@di.ubi.pt

I. M. Pires · G. Marques · N. M. Garcia
Instituto de Telecomunicações, Universidade da Beira Interior, Covilhã, Portugal

G. Marques
Instituto Politécnico da Guarda, Guarda, Portugal

E. Zdravevski · P. Lameski
Faculty of Computer Science and Engineering, University Ss Cyril and Methodius, Skopje, Macedonia
e-mail: eftim.zdravevski@finki.ukim.mk

P. Lameski
e-mail: petre.lameski@finki.ukim.mk

F. Flórez-Revuelta
Department of Computer Technology, Universidad de Alicante, Alicante, Spain
e-mail: francisco.florez@ua.es

S. Spinsante
Department of Information Engineering, Marche Polytechnic University, Ancona, Italy
e-mail: s.spinsante@univpm.it

V. E. Balas et al. (eds.), *Internet of Things and Big Data Applications*, Intelligent Systems Reference Library 180, https://doi.org/10.1007/978-3-030-39119-5_7

This paper focuses on the review of the mobile applications available in the Google Play Store that are dedicated to cardiac patients. The number of cardiac patients is increasing, but there are no mobile applications that aid cardiac patients by providing monitoring of different parameters, including the calorie intake and the calories burned. However, the mobile applications that can be adapted to this type of people were analyzed. We found six notable mobile applications. Their features can be grouped in diet, anthropometric parameters, and physical activity.

Keywords Ambient Assisted Living · Cardiac patients · Diet · Enhanced living environments · Physical exercise · Mobile computing · Mobile health

7.1 Introduction

The number of people with heart rate related problems is growing [1, 2]. However, technology may help to control the symptoms and improve the quality of life. The development of mobile applications that provide specific exercises and measure calories burned and intake may help in control and minimize the heart rate problems [3–6].

Ambient Assisted Living is a multi-disciplinary approach to purpose innovative telehealth and personalized healthcare systems by combining microcontrollers, sensors, actuators, wireless networks, mobile computing technologies, and open-source software on the same platform for enhanced living environments [7–10]. Ambient Assisted Living is closely related to information and communication technologies to design and develop appropriate healthcare supervision to promote health and well-being for older adults [11]. The enhanced living environment is a concept closely related to the Ambient Assisted Living field. However, enhanced living environments are more associated with information and communications technologies than Ambient Assisted Living [12, 13]. Enhanced living environments include the more recent information and communications technologies achievements to support Ambient Assisted Living systems for enhanced personalized healthcare. Furthermore, Enhanced living environments incorporate information and communications technologies such as algorithms, platforms, and systems to design and develop innovative applications and services. These systems aim to maintain an independent and autonomous living of older adults for as long as possible. Likewise, enhanced living environments include the latest technological achievements related to the Internet of Things to create intelligent systems for enhanced people's health and well-being. A personalized healthcare solution is a set of hardware and software systems developed to promote a wide range of medical services to all the individuals in general and older adults in particular. The healthcare systems incorporate mobile computing technologies and practical and often pervasive methods to promote health and well-being for both medical staff and patients. These systems are closely related to enhanced living environments and Ambient Assisted Living.

Furthermore, they propose an efficient and effective potential to address several healthcare issues through the incorporation of mobile computing technologies and medical systems. An enhanced living environment incorporates an ecosystem of medical systems, which include medical sensors, microcontrollers, wireless communication technologies, and open-source software platforms for data visualization and analytics. This ecosystem incorporates different monitoring solutions that make use of open source technologies for data acquisition, transmission, and processing microsensors. These environmental monitoring solutions provide secure and simultaneous access to a database collected from different sites using mobile computing technologies for different proposes such as air quality monitoring, noise monitoring, activity, environment recognition, and thermal and light comfort assessment [14–29]. Moreover, recent Ambient Assisted Living systems must incorporate mobile computing technologies and several sensors for measuring position, location, oxygen, blood pressure, temperature, glucose, and weight. Most of these systems incorporate wireless communications technologies such as Ethernet, Bluetooth, ZigBee, and Wi-Fi.

It is anticipated that in 2050, 20% of the total world population will be aged 60 years or older [30]. That will lead to a relevant proliferation of disorders, medical services costs, and a lack of caregivers. Furthermore, the Ambient Assisted Living research field is also closely related to mobile health. Mobile health is a research area that aims to provide personalized medical services through mobile computing technologies using a cost-effective approach on a real-time basis for enhanced public health [31, 32]. These medical services and applications will support and improve the quality of the integration between patient and doctor, which will be relevant for enhanced personalized healthcare systems [33–35]. The incorporation of mobile computing technologies in personalized healthcare systems for enhanced health and well-being is a trending topic. There is an active proliferation of interest in the research and design of efficient systems that incorporate mobile computing technologies for enhanced living environments [34].

Nevertheless, most individuals do not use mobile applications to promote their health, and the people who use this kind of solution tend to stop using them in a short period. Consequently, the researchers must focus on new efficient and effective pervasive and ubiquitous methods to address the users' requirements to promote the adoption of these systems and improve the mobile application quality [36]. In the recent past, the processing power of mobile devices and the evolution in the wireless communication technologies field provide the requirements to design and develop mobile health systems, which lead to the relevant evolution of the digitalization of medical services and personalized healthcare [37]. Ambient Assisted Living systems and mobile computing will address most of the challenges of older adults adequately and will extend their independence [38–40]. This is a relevant topic as people prefer to remain in their homes and support the high cost related to nursing care and the installation of innovative healthcare systems instead of been transferred to nursing homes [41]. The continuous technological enhancements provide the requirements for the design and development of real-time monitoring systems that provide automatic and innovative personalized healthcare services to promote the creation of enhanced living environments and improve public health [42–44]. These technological advancements in different areas such as networking, microcontrollers, sensors, provide cost-effective monitoring to support real-time patient monitoring which leads to the creation of enhanced living environments [45].

Even though the numerous open-issues in the research and development of pervasive and ubiquitous Ambient Assisted Living and personalized healthcare systems, such as data architecture, human-computer interaction, ergonomics, and accessibility these systems continue to provide relevant medical activities every day [46]. Furthermore, numerous social and ethical challenges are reported in the literature. These challenges include the acceptance by the older adults and the security problems which are a relevant requirement for the design and implementation of these personalized healthcare systems.

It must be noted that it is essential to mention that personal care cannot be replaced with technology; Ambient Assisted Living systems can be considered as an essential addition to an independent lifestyle [47]. Furthermore, these systems can directly or indirectly help to maintain older adults within their home environments in conjunction with personal care instead of been moved into institutionalized environments.

The development and review of mobile applications related to cardiac patients are included in the development of Ambient Assisted Living systems [7, 10, 12, 13, 48–50]. Cardiac patients have a limited exercise capacity, so they need specialized exercise for them. Most of these exercises are recommended individually by doctors, so it is hard to find exercises that fit all kinds of cardiac problems [51, 52]. This paper focuses on the review of mobile applications to calorie intake control and specific exercises designed for cardiac patients. These applications must include recommended practices and food for the users' specific conditions.

The motivation for this paper is to identify the features of these types of mobile applications available in the market for the creation of a novel mobile application that can help people with cardiac-related problems to lead a normal healthy life coupled with exercise indicated for them. In order to develop the mobile application, we are going to study already available mobile applications similar to the subject of this paper. Next, we are going to use the information obtained to develop a mobile application with the combination of the useful parts of the already present application and the missing features.

This paper is going to feature a review of similar mobile applications based on selection criteria and evaluate their usefulness for cardiac patients and their functionalities, dividing the features among diet, anthropometric parameters, and physical activity. This paragraph ends the introduction section. Section 7.2 presents the methodology for this review, describing the results in Sect. 7.3. A taxonomy is proposed, and the results are discussed in Sect. 7.4. Finally, Sect. 7.5 concludes this review.

7.2 Methodology

7.2.1 Research Questions

The questions asked for this research where: (RQ1) What Mobile Applications available for calorie intake control to help on a diet?; (RQ2) Which of these Mobile Applications can help someone with cardiac problems lead a healthy life based on exercise and Calorie intake control?; (RQ3) How can we group the functionalities of the different applications.

7.2.2 Inclusion Criteria

The applications included in this paper follow the following inclusion criteria: (1) Applications related with calories control and exercise; (2) must be in English or Portuguese; (3) the application is free; (4) it was updated in 2018 or later; (5) application must have more than 10,000 downloads.

7.2.3 Search Strategy

In this study, we used the following combination of keywords: "physical activity" "cardiac" and "diet". This was used to identify applications with a similar theme. Afterward, the mobile applications were analyzed according to the inclusion and exclusion criteria based on the characteristics and functionality.

7.2.4 Extraction of Study Characteristics

After analyzing each one of the applications, the following information was extracted (see Table 7.1): name, user rating, number of downloads at the date of analysis, author, use of sensors, presence in studies, and year of the last update.

7.3 Methodology

The search identified 116 mobile applications, as illustrated in Fig. 7.1. The mobile applications were evaluated in terms of user rating, the number of downloads, the use of sensors, presence in scientific studies, and year of the last update, all these evaluations are at the time of the review.

Table 7.1 List of mobile applications analyzed

Name	User rating	Number of downloads	Author	Use of sensors	Scientific studies	Year of the last update
iEatBetter: Food Diary [53]	4.1	1,000,000	My daily bits LLC	No	No	2019
Calorie counter by FatSecret [54]	4.5	10,000,000	FatSecret	Yes	Yes	2018
Calorie counter—MyFitnessPal [55]	4.6	50,000,000	MyFitnessPal, Inc.	Yes	Yes	2019
YAZIO Calorie counter [56]	4.6	500,000	YAZIO	Yes	Yes	2019
ProHealth Diet [57]	–	10	CoachCare	Yes	No	2018
My Diet Coach [58]	4.4	10,000,000	Health and fitness	No	Yes	2018

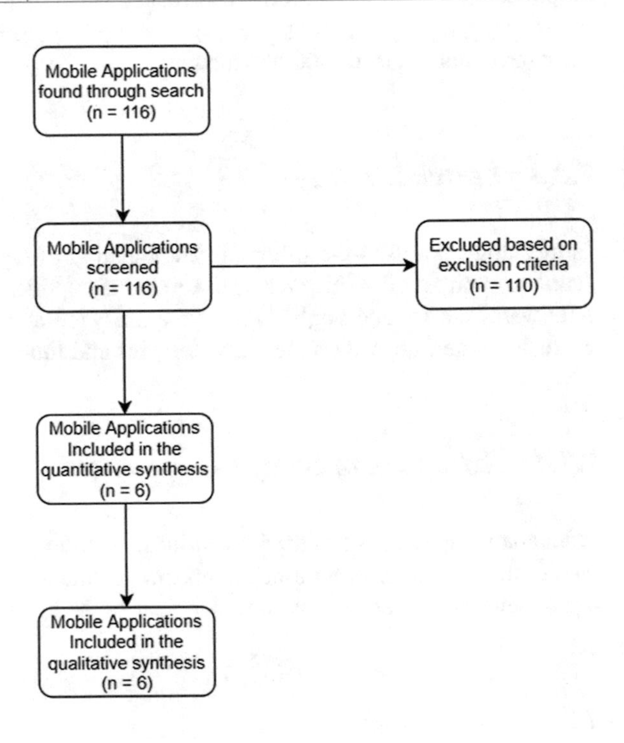

Fig. 7.1 Mobile application analysis

The mobile applications were first selected based on the exclusion criteria, which eliminate 110 applications since most of them were quizzes for students to learn about the cardiac system and various related health problems. The inclusion criteria did not eliminate any mobile applications from the review since they were already met. The remaining six mobile applications were included in the review.

Table 7.1 shows User rating, number of downloads, use of sensors, presence in scientific studies, and year of the last update. Following the User rating, there are 1 mobile application with 4.1 stars (16.7%), 1 mobile application with 4.4 stars (16.7%), 1 mobile application with 4.5 stars (16.7%), 2 mobile applications with 4.6 stars (33.3%) and 1 mobile application without reviews (16.7%). Following the number of downloads, there are 1 mobile application with at least 10 downloads (16.7%), 1 mobile application with at least 500,000 downloads (16.7%), 1 mobile application with at least 1,000,000 downloads (16.7%), 2 mobile applications with at least 10,000,000 downloads (33.3%) and 1 mobile application with 50,000,000 downloads (16.7%). The sensors available in the mobile devices are used in 4 mobile applications (66.7%). Following the year of the last update, there are three mobile applications updated in 2018 (50.0%) and three mobile applications updated in 2019 (50.0%).

iEatBetter: Food Diary [53], an app with no login needed to keep track of what we eat, helping lose maintain or gain weight. It has a database with items included and keeps a journal of digested food with calories, carbs, fats, and protein.

Calorie Counter by FatSecret [54] is an app to track food intake, ability to use image recognition of food and barcode scanner, is also able to record all calories burned through exercise and supports the integration of tracking applications.

Calorie Counter—MyFitnessPal [55], provide health and fitness tracking applications that support barcode scanning, recipes, and restaurant menus. It can sync with other applications and track steps.

YAZIO Calorie Counter, Nutrition Diary & Diet Plan [56] Application aims to lose weight and build muscle based on calories, carbs, proteins, and fats tracking. Tracks sports and exercises. Has a built-in barcode scanner and has food recipes with included calorie information. Syncs with Google fit.

ProHealth Diet [57] provides a Doctor-supervised weight loss program, helps people reach their weight goal and maintain it.

My Diet Coach—Weight Loss Motivation & Tracker [58] Application focuses on motivation and inspiration to lose weight, it has tips and a reminder to help the user. It has a water consumption tracker.

7.4 Discussion

In our research, we found there is a lack of scientific papers featuring reviews of calorie and exercise control applications dedicated to cardiac patients, the reason being nonexistence of these applications, so we cannot compare the results of this paper to the results of others.

The main features of these applications are:

- Diet—These applications provide a database with various foods and their calories, which the user can select and a diary to keep track of the food eaten in each day. This is used to calculate calorie intake.
- Anthropometric parameters—These applications collect the age, weight, height, and gender of the users. This data is essential to calculate the basal metabolic rate.
- Physical Activity—These applications can track the calories burned during physical exercise, the user can select the exercise performed, and the application calculates calories burned based on the time a given activity was performed. These applications also have a database with exercises and their calorie-burning rate. The applications also keep track of the exercises performed every day.

Based on the functionalities of these applications, we propose the categorization of the main features of the mobile applications researched in Fig. 7.2. However, all of these mobile applications are very similar.

Cardiac patients require a specific type of exercise specially tailored for their condition and physical capabilities. In general, these patients have a very limited physical capacity that can be improved with low to moderate intensity exercises [51, 52].

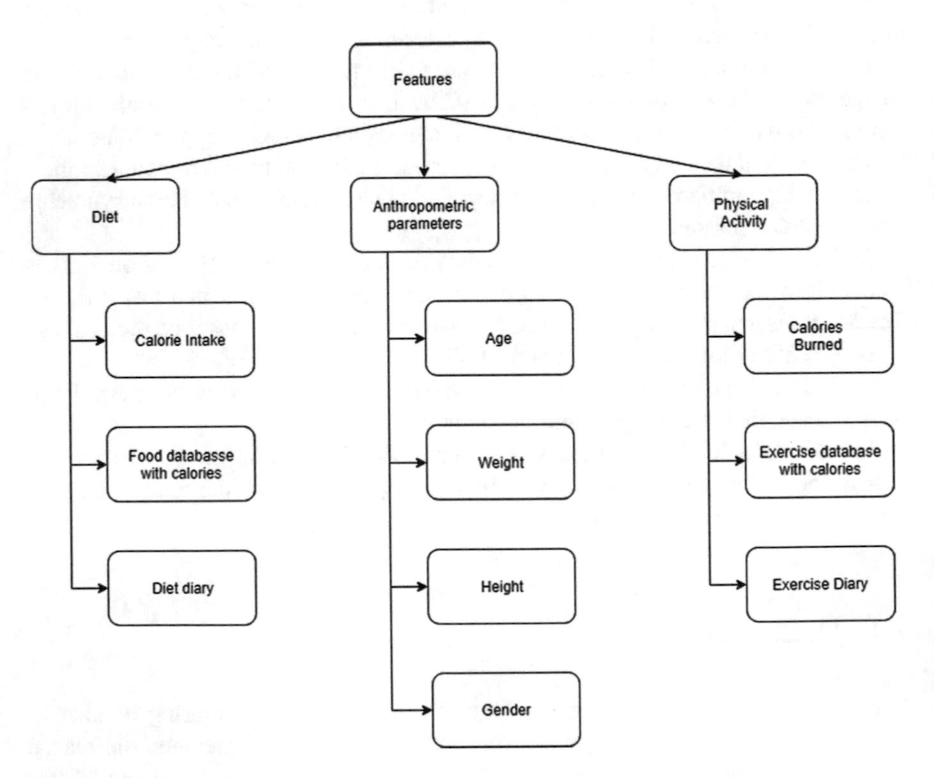

Fig. 7.2 Mobile applications taxonomy

Of the reviewed applications, the following were present in articles: Calorie Counter by FatSecret [35, 54, 59];Calorie Counter—MyFitnessPal [35, 55, 59]; YAZIO Calorie Counter, Nutrition Diary & Diet Plan [56, 60, 61]; and My Diet Coach—Weight Loss Motivation & Tracker [58, 59, 61, 62]. In these studies, these mobile applications are reviewed in detail, and in some of the articles, they are compared with each other.

Today mobile devices are used by most people in their daily routine. Therefore the development of mobile applications for multiple purposes in general and health support, in particular, is relevant for enhanced public health [63]. Furthermore, mobile developers must focus on persuasive principles, particularly for mobile health applications, due to the relevant contribution of these new methods to promote public health and well-being. Several studies report that mobile health applications have a positive impact on health and well-being as these applications contribute to encouraging users to be more actively involved in their healthcare management. Moreover, mobile applications propose an effective method to encourage physical activities [64].

The proliferation of mobile computing technologies has revolutionized the medical practices in both patient and clinical staff sides. The development of mobile health applications continues to increase; mainly, the cardiology field is the most addressed [65]. The mobile applications play a significant role in promoting health as people are getting more attention to their health and wellness through mobile computing technologies. The medical decision-making systems are relevant for the daily routine of clinical staff, and the information and communication technologies provide enhanced methods for patient data consulting and help the clinical staff to manage their tasks. Soon, the digitalization of healthcare systems through mobile applications will continue to be addressed by several research fields to provide new architectures and methods to consult and manage electronic patient data, electronic prescriptions, and medical notes [66, 67].

7.5 Conclusions

This review concluded the following about the researched mobile applications. This paper answers the following research question that this paper is based on:

- (RQ1) All the applications screened are capable of helping someone lead a healthy life based on calorie control. Every single one of them follows the same principle, but with different functionalities. Some can use sensor and sync with other apps. This approach makes it easier to calculate the calories burned during the day.
- (RQ2) Of the screened applications, none of them was specially dedicated to the use of patients of cardiac problems. This shows a significant void in the market for applications for these patients.
- (RQ3) The functionalities of these mobile applications are generally grouped in Diet, Physical Activity, and Anthropometric parameters.

This study concludes that most of the applications screened in this paper are very similar. The primary purpose is the control of weight through calorie intake management. From all the applications, 2/3 can use sensors to determine the physical activity of the user. From the applications screened, 50% have a barcode scanner for food.

The screened applications are capable of calculating the calorie intake through the selection of the food eaten within the application. They are also capable of calculating the burned calories through the selection of the physical exercise performed.

Some of these applications, given the broad approach of the subject of exercise and calorie consumption, can be used by cardiac patients but can lead to misinformation given the type of exercises that these people should not perform.

Of the total of six mobile applications, four are featured in scientific papers in which some of them are compared with other applications. However, the main conclusion of this paper is that there is a severe lack of applications dedicated to cardiac patients.

In general, these mobile applications are only used in scientific studies, but they are not available in the online application stores.

Acknowledgements This work is funded by FCT/MEC through national funds and when applicable co-funded by FEDER—PT2020 partnership agreement under the project UID/EEA/50008/2019 (Este trabalho é financiado pela FCT/MEC através de fundos nacionais e quando aplicável cofinanciado pelo FEDER, no âmbito do Acordo de Parceria PT2020 no âmbito do projeto UID/EEA/50008/2019). This article is based upon work from COST Action IC1303—AAPELE—Architectures, Algorithms and Protocols for Enhanced Living Environments and COST Action CA16226—SHELD-ON—Indoor living space improvement: Smart Habitat for the Elderly, supported by COST (European Cooperation in Science and Technology). More information in www.cost.eu.

References

1. Gaziano, T.A., Bitton, A., Anand, S., Abrahams-Gessel, S., Murphy, A.: Growing epidemic of coronary heart disease in low- and middle-income countries. Curr. Probl. Cardiol. **35**(2), 72–115 (2010)
2. Supino, P.G., Borer, J.S., Yin, A., Dillingham, E., McClymont, W.: The epidemiology of valvular heart diseases: the problem is growing. In: Borer, J.S., Isom, O.W. (eds.) Advances in Cardiology, vol. 41, pp. 9–15. KARGER, Basel (2004)
3. Gay, V., Leijdekkers, P., Barin, E.: A mobile rehabilitation application for the remote monitoring of cardiac patients after a heart attack or a coronary bypass surgery. In: Proceedings of the 2nd International Conference on Pervasive Technologies Related to Assistive Environments—PETRA '09, pp. 1–7. Corfu, Greece (2009)
4. Qudah, I., Leijdekkers, P., Gay, V.: Using mobile phones to improve medication compliance and awareness for cardiac patients. In: Proceedings of the 3rd International Conference on Pervasive Technologies Related to Assistive Environments—PETRA '10, p. 1. Samos, Greece (2010)
5. Beatty, A.L., Fukuoka, Y., Whooley, M.A.: Using mobile technology for cardiac rehabilitation: a review and framework for development and evaluation. J. Am. Heart Assoc. **2**(6) 2013

6. Kakria, P., Tripathi, N.K., Kitipawang, P.: A real-time health monitoring system for remote cardiac patients using smartphone and wearable sensors. Int. J. Telemed. Appl. **2015**, 1–11 (2015)
7. Garcia, N.M.: A roadmap to the design of a personal digital life coach. In: Loshkovska, S., Koceski, S. (eds.) ICT Innovations 2015, vol. 399, pp. 21–27. Springer International Publishing, Cham (2016)
8. Ganchev, I., Garcia, N.M., Dobre, C., Mavromoustakis, C.X., Goleva, R. (eds.): Enhanced Living Environments: Algorithms, Architectures, Platforms, and Systems, vol. 11369. Springer International Publishing, Cham (2019)
9. Mavromoustakis, C.X., Garcia, N.M., Goleva, R.I., Mastorakis, G., Dobre, C. (eds.): Ambient assisted living and enhanced living environments: principles, technologies and control. Butterworth-Heinemann, Amsterdam; Boston (2017)
10. Garcia, N.M., Rodrigues, J.J.P.C. (eds.): Ambient Assisted Living, 0 edn. CRC Press (2015)
11. Marques, G., Pitarma, R., Garcia, N.M., Pombo, N.: Internet of things architectures, technologies, applications, challenges, and future directions for enhanced living environments and healthcare systems: a review. Electronics **8**(10), 1081 (2019)
12. Dobre, C., Mavromoustakis, C.X., Garcia, N.M., Mastorakis, G., Goleva, R.I.: Introduction to the AAL and ELE Systems. In: Ambient Assisted Living and Enhanced Living Environments, pp. 1–16. Elsevier (2017)
13. Goleva, R.I., et al.: AAL and ELE platform architecture. In: Ambient Assisted Living and Enhanced Living Environments, pp. 171–209. Elsevier (2017)
14. Marques, G., Pitarma, R.: Monitoring and control of the indoor environment. In: 2017 12th Iberian Conference on Information Systems and Technologies (CISTI), pp. 1–6 (2017)
15. Pitarma, R., Marques, G., Ferreira, B.R.: Monitoring indoor air quality for enhanced occupational health. J. Med. Syst. **41**(2) (2017)
16. Marques, G., Pitarma, R.: Health informatics for indoor air quality monitoring. In: 2016 11th Iberian Conference on Information Systems and Technologies (CISTI), pp. 1–6 (2016)
17. Marques, G., Roque Ferreira, C., Pitarma, R.: A system based on the internet of things for real-time particle monitoring in buildings. Int. J. Environ. Res. Public Health **15**(4), 821 (2018)
18. Feria, F., Salcedo Parra, O.J., Reyes Daza, B.S.: Design of an architecture for medical applications in IoT. In: Luo, Y. (ed.) Cooperative Design, Visualization, and Engineering, vol. 9929, pp. 263–270. Springer International Publishing, Cham (2016)
19. Ray, P.P.: Internet of things for smart agriculture: technologies, practices and future direction. J. Ambient Intell. Smart Environ. **9**(4), 395–420 (2017)
20. Matz, J.R., Wylie, S., Kriesky, J.: Participatory air monitoring in the midst of uncertainty: residents' experiences with the speck sensor. Engag. Sci. Technol. Soc. **3**, 464 (2017)
21. Demuth, D., Nuest, D., Bröring, A., Pebesma, E.: The airquality sensebox. In: EGU General Assembly Conference Abstracts, vol. 15 (2013)
22. Marques, G., Pitarma, R.: A cost-effective air quality supervision solution for enhanced living environments through the internet of things. Electronics **8**(2), 170 (2019)
23. Marques, G., Ferreira, C.R., Pitarma, R.: Indoor air quality assessment using a CO_2 monitoring system based on internet of things. J. Med. Syst. **43**(3) 2019
24. Marques, G., Pitarma, R.: Monitoring and control of the indoor environment. In: 2017 12th Iberian Conference on Information Systems and Technologies (CISTI), pp. 1–6. Lisbon, Portugal (2017)
25. Marques, G., Pitarma, R.: mHealth: indoor environmental quality measuring system for enhanced health and well-being based on internet of things. J. Sens. Actuator Netw. **8**(3), 43 (2019)
26. Marques, G., Pitarma, R.: Noise monitoring for enhanced living environments based on internet of things. In: Rocha, Á., Adeli, H., Reis, L.P., Costanzo, S. (eds.) New Knowledge in Information Systems and Technologies, vol. 932, pp. 45–54. Springer International Publishing, Cham (2019)
27. Marques, G., Pitarma, R.: Noise mapping through mobile crowdsourcing for enhanced living environments. In: Rodrigues, J.M.F., Cardoso, P.J.S., Monteiro, J., Lam, R., Krzhizhanovskaya, V.V., Lees, M.H., Dongarra, J.J., Sloot, P.M.A. (eds.) Computational Science—ICCS 2019, vol. 11538, pp. 670–679. Springer International Publishing, Cham (2019)

28. Marques, G., Pitarma, R.: Air quality through automated mobile sensing and wireless sensor networks for enhanced living environments. In: 2019 14th Iberian Conference on Information Systems and Technologies (CISTI), pp. 1–7. Coimbra, Portugal (2019)
29. Pires, I.M., Garcia, N.M., Pombo, N., Flórez-Revuelta, F., Spinsante, S., Teixeira, M.C.: Identification of activities of daily living through data fusion on motion and magnetic sensors embedded on mobile devices. Pervasive Mobile Comput. **47**, 78–93 (2018)
30. UN,'Worldpopulationageing:1950–2050', pp. 11–13 (2001)
31. Huh, J.-H., Kim, T.-J.: A location-based mobile health care facility search system for senior citizens. J. Supercomput. **75**(4), 1831–1848 (2019)
32. Villani, G.Q., et al.: Mobile health and implantable cardiac devices: patients' expectations. Eur. J. Prev. Cardiolog. **26**(9), 920–927 (2019)
33. Silva, B.M.C., Rodrigues, J.J.P.C., de la Torre Díez, I., López-Coronado, M., Saleem, K.: Mobile-health: a review of current state in 2015. J. Biomed. Informatics **56**, 265–272 (2015)
34. Stoyanov, S.R., Hides, L., Kavanagh, D.J., Zelenko, O., Tjondronegoro, D., Mani, M.: Mobile app rating scale: a new tool for assessing the quality of health mobile apps. JMIR mHealth uHealth **3**(1), e27 (2015)
35. Azfar, A., Choo, K.-K.R., Liu, L.: Forensic Taxonomy of Popular Android mHealth Apps. arXiv: 1505.02905 [cs] (2015)
36. Krebs, P., Duncan, D.T.: Health app use among us mobile phone owners: a national survey. JMIR mHealth uHealth **3**(4), e101 (2015)
37. Steinhubl, S.R., Muse, E.D., Topol, E.J.: The emerging field of mobile health. Sci. Transl. Med. **7**(283), 283rv3–283rv3 (2015)
38. Moumtzoglou, A. (ed.): Mobile Health Applications for Quality Healthcare Delivery. IGI Global (2019)
39. Morita, P.P., et al.: A patient-centered mobile health system that supports asthma self-management (breathe): design, development, and utilization. JMIR mhealth uhealth **7**(1), e10956 (2019)
40. Buckingham, S.A., Williams, A.J., Morrissey, K., Price, L., Harrison, J.: Mobile health interventions to promote physical activity and reduce sedentary behaviour in the workplace: a systematic review. Digital Health **5**, 205520761983988 (2019)
41. Centers for Disease Control and Prevention: The state of aging and health in America 2007, N. A. on an Aging Society. Available: https://www.cdc.gov/aging/pdf/saha_2007.pdf (2007)
42. Manogaran, G., Chilamkurti, N., Hsu, C.-H.: Emerging trends, issues, and challenges in internet of medical things and wireless networks. Personal Ubiquitous Comput. (2018)
43. Manogaran, G., Varatharajan, R., Lopez, D., Kumar, P.M., Sundarasekar, R., Thota, C.: A new architecture of internet of things and big data ecosystem for secured smart healthcare monitoring and alerting system. Future Generation Comput. Syst. **82**, 375–387 (2018)
44. Rathore, M.M., Paul, A., Ahmad, A., Chilamkurti, N., Hong, W.-H., Seo, H.: Real-time secure communication for smart city in high-speed big data environment. Future Generation Comput. Syst. **83**, 638–652 (2018)
45. Jo, D., Kim, G.J.: ARIoT: scalable augmented reality framework for interacting with internet of things appliances everywhere. IEEE Trans. Consum. Electron. **62**(3), 334–340 (2016)
46. Koleva, P., Tonchev, K., Balabanov, G., Manolova, A., Poulkov, V.: Challenges in designing and implementation of an effective ambient assisted living system. In: 2015 12th International Conference on Telecommunication in Modern Satellite, Cable and Broadcasting Services (TELSIKS), pp. 305–308 (2015)
47. Marques, G., Pitarma, R.: IAQ evaluation using an IoT CO_2 monitoring system for enhanced living environments. In: Rocha, Á., Adeli, H., Reis, L.P., Costanzo, S. (eds.) Trends and Advances in Information Systems and Technologies, vol. 746, pp. 1169–1177. Springer International Publishing, Cham (2018)
48. Pires, I., Garcia, N., Pombo, N., Flórez-Revuelta, F.: From data acquisition to data fusion: a comprehensive review and a roadmap for the identification of activities of daily living using mobile devices. Sensors **16**(2), 184 (2016)

49. Sousa, P.S., Sabugueiro, D., Felizardo, V., Couto, R., Pires, I., Garcia, N.M.: mHealth sensors and applications for personal aid. In: Adibi, S. (ed.) Mobile Health, vol. 5, pp. 265–281. Springer International Publishing, Cham (2015)
50. Dolićanin, Ć., Kajan, E., Randjelović, D., Stojanović, B. (eds.): Handbook of Research on Democratic Strategies and Citizen-Centered E-Government Services. IGI Global (2015)
51. Franklin, B.A., Gordon, S., Timmis, G.C.: Amount of exercise necessary for the patient with coronary artery disease. Am. J. Cardiology **69**(17), 1426–1432 (1992)
52. Franklin, B.A.: How much exercise is enough for the coronary patient? Preventive Cardiology **3**(2), 63–70 (2000)
53. iEatBetter:FoodDiary. https://play.google.com/store/apps/details?id=com.dailybits.foodjournal. (2019)
54. CalorieCounterbyFatSecret. [Online]. Available: https://play.google.com/store/apps/details?id=com.fatsecret.android (2019)
55. CalorieCounter-MyFitnessPal. [Online]. Available: https://play.google.com/store/apps/details?id=com.myfitnesspal.android&hl=en (2019)
56. YAZIO Calorie Counter. [Online]. Available: https://play.google.com/store/apps/details?id=com.yazio.android (2019)
57. ProHealthDiet. [Online]. Available: https://play.google.com/store/apps/details?id=com.coachcare.prohealthdietionic (2019)
58. My Diet Coach.: [Online]. Available: https://play.google.com/store/apps/details?id=com.dietcoacher.sos (2019)
59. Franco, R.Z., Fallaize, R., Lovegrove, J.A., Hwang, F.: Popular nutrition-related mobile apps: a feature assessment. JMIR mHealth uHealth **4**(3), e85 (2016)
60. Orso, V., Spagnolli, A., Viero, F., Gamberini, L.: The design, implementation and evaluation of a mobile app for supporting older adults in the monitoring of food intake. In: Leone, A., Caroppo, A., Rescio, G., Diraco, G., Siciliano, P. (eds.) Ambient Assisted Living, vol. 544, pp. 147–159. Springer International Publishing, Cham (2019)
61. Darby, A., Strum, M.W., Holmes, E., Gatwood, J.: A review of nutritional tracking mobile applications for diabetes patient use. Diabetes Technol. Ther. **18**(3), 200–212 (2016)
62. Chen, J., Cade, J.E., Allman-Farinelli, M.: The most popular smartphone apps for weight loss: a quality assessment. JMIR mHealth uHealth **3**(4), e104 (2015)
63. Li, H., Shou, G., Hu, Y., Guo, Z.: Mobile edge computing: progress and challenges. In: 2016 4th IEEE International Conference on Mobile Cloud Computing, Services, and Engineering (MobileCloud), pp. 83–84. Oxford, United Kingdom (2016)
64. Matthews, J., Win, K.T., Oinas-Kukkonen, H., Freeman, M.: Persuasive technology in mobile applications promoting physical activity: a systematic review. J. Med. Syst. **40**(3), 72 (2016)
65. de Garibay, V.G., Fernández, M.A., de la Torre-Díez, I., López-Coronado, M.: Utility of a mHealth app for self-management and education of cardiac diseases in spanish urban and rural areas. J. Med. Syst. **40**(8), 186 (2016)
66. Costa, S.E.P., Rodrigues, J.J.P.C., Silva, B.M.C., Isento, J.N., Corchado, J.M.: Integration of wearable solutions in AAL environments with mobility support. J. Med. Syst. **39**(12), 184 (2015)
67. Bhuyan, S.S., et al.: Use of mobile health applications for health-seeking behavior among US adults. J. Med. Syst. **40**(6), 153 (2016)

Chapter 8
Review on Cyber Security Intrusion Detection: Using Methods of Machine Learning and Data Mining

Rajshree Sriavstava, Pawan Singh and Hargun Chhabra

Abstract Today internet has become a crucial part of our day to day life. Communication, banking, travelling, shopping and learning are done using Internet. A system that is not properly secured will get infected and their vital information can be acquired using hacking. Cyber security is a method using which our data can be secured, but even this has its downfall, as it can't predict forthcoming new attacks. Using ML and DM techniques we can detect those intrusions and stop them from occurring. This review contains information about how Intrusion detection methods in cyber security can be used with Machine Learning and Data Mining techniques.

Keywords Internet · Secured · Cyber security · Downfall · Forthcoming · ML · DM · Intrusion · Detection

8.1 Introduction

The arrival of networking, communication and information has created a new world of densely populated users connected to each other in a cyber space. This network enables them to set up link between each other and share their information.

Cyber security is a method that enables the users to be ensured that their valuable data is not beaching out. It blocks all the malicious activities and malwares that attempt to force their way in a network and misuse the data that is being secured by a user. Inventors of cyber security Bob Thomas realize that a computer program can move across a network, leaving a mark wherever it went. He created—Creeper to travel between Tenex terminals on Arpanet, printing—I'M THE CREEPER-CATCH-ME-IF-YOU CAN.

R. Sriavstava (✉) · P. Singh (✉) · H. Chhabra (✉)
Department of Computer Science and Engineering, DIT University, Dehradun, Uttrakhand, India
e-mail: rajshree.srivastava27@gmail.com

P. Singh
e-mail: pawan.bisht1809@gmail.com

H. Chhabra
e-mail: hargunyo@gmail.com

V. E. Balas et al. (eds.), *Internet of Things and Big Data Applications*, Intelligent Systems Reference Library 180, https://doi.org/10.1007/978-3-030-39119-5_8

This lead to some more mischief activities until in 1986, Marcus Hess hacked an internet gateway in Berkeley, 400 military computers got hacked (including mainframe at Pentagon), with intent to sell them to KGB. Luckily Clifford Stoll detected intrusion and deployed honeypot technique. Since then computer virus was considered as serious threats not just academic pranks. Some of latest cyber security tools are Metasploit Framework, Nmap, Wireshark, Nessus etc.

Data Mining (DM) is a term that came to be known in 1990s after the Knowledge Discovery in Databases, which was conceived by Piatetsky—Shapiro. Data mining is a method in which we analyse and gather useful information from large amount of data. Few of Data Mining techniques are Classification analysis, Association rule learning, Anomaly detection, Clustering analysis, Regression analysis.

Intrusion Detection System (IDS) are devices or software's that monitor malicious activity and report about them. Types of IDS are Network Intrusion Detection System (NIDS), Host-based Intrusion Detection System (HIDS). Signature based Intrusion Detection System, Anomaly-based Intrusion Detection System. Intrusion Detection and Prevention System (IDPS). These are applications that are used to detect a malicious activity, make a log about it, report that activity and attempt to block or stop it. Four types of IDPS are Network-based Intrusion Prevention System (NIPS), Wireless Intrusion Prevention System and Host-based Intrusion Prevention System (HIPS).

The paper is being divided into sections namely Sect. 8.1 represents Introduction, Sect. 8.2 machine learning and data mining concept, Sect. 8.3 importance of data sets for ML and DM, Sect. 8.4 ML and DL methods for intrusion detection, Sect. 8.5 Conclusion and Sect. 8.6 future scope.

8.2 Machine Learning and Data Mining Concept

Machine Learning is the subset of Artificial intelligence. The name machine learning (ML) was coined in 1959 by Arthur Samuel. He wrote the first computer learning program. The program was the game of checkers and through which the history of machine learning evolved.

Machine Learning is defined as the study and construction of algorithms that can learn from and make predictions on data sets.
There are two types of Machine Learning techniques

1. Supervised Learning
2. Unsupervised Learning

A Supervised learning algorithm analyses given data sets and produces a generalized function, which can be used for mapping inputs to desired output. e.g. classifying email as spam email. It is combination of both classification and regression techniques to develop prediction. Classifications predict discrete responses and classify labelleddata. Regression techniques predict continuous responses using previous labelled data.

Common algorithms for performing classification and regression are:

Classification Algorithms: Support Vector machine (SVM), boosted and bagged decision trees, Naïve Bayes, logistic regression and neural network.
Regression Algorithms: Linear model, nonlinear model etc.
Unsupervised Learning: uses machine learning algorithms that draw conclusions on unlabelled data without having any predefined dataset for its training.

Clustering is the most common unsupervised learning techniques. Generally, this technique is used for exploratory data analysis to find hidden patterns or groupings in data. Common algorithms for Clustering techniques: k-means, fuzzy c-means clustering etc.

8.3 Importance of Datasets for ML and DM

In this section we are going to discuss about data sets in details for ML and DM methods or techniques. It is really important as these methods learn from the data available in the data sets. This section will give us a full idea of different type of data used by Machine Learning and Data Mining methods: packet capture, network data (net flow), and public data set. Detailed description of algorithms in DM and ML methods will be given in Sect. 8.4.

It is important to understand what Data sets are. Data sets are nothing but collection of data which may be in organized or unorganized manner from which ML and DM methods learn. Types of learning are discussed in Sect. 8.2.

There are many protocols that we use in our day to day life which generates packet, some of these protocols are as Internet Protocol Version-4, Internet Protocol Version-6, User Datagram Protocol, Internet Control Message Protocol, Transmission Control Protocol, and Internet Gateway Management Protocol. These packets can be captured at Ethernet ports by applications like pcap. libcap (Unix based library), wincap (windows based library). These libraries are used in many applications like Tcp dump and Wire Shark [1] etc.

In network physical layer there is an Ethernet frame which consist of header, Ethernet header is actually the MAC address and it consist of a 1500 byte of payload, the payload consists of IP packet. IP packets consists of an IP header and IP payload. IP payload consist of other protocols (the process of one protocol consisting of another protocol is called encapsulation). Some of these protocols are (HTTP), File Transfer Protocol (FTP), and Post Office Protocol (POP) [1, 2] etc.

As the packet captured by the pcap interface has different protocols, so data also varies based on protocol contained in the packet. Data Flow is nothing but a method by which we design or map the data of IP traffic, many specialized softwares are used to map the data flow [1].

Net-Flow Data is a protocol developed by Cisco which Analyses the flow of IP traffic using Net-Flow Collector and analyser we can see from where is the traffic. Coming from. Other similar technologies are:

- Juniper (J-Flow)
- 3Com/HP, Dell, and Netgear (s-flow)
- Huawei (NetStream)
- Alcatel-Lucent (Cflowd)
- Ericsson (RFlow)

Public Data **sets** helps in foundation of those who are unfamiliar to big data and data analysis and is considered as a powerful repository for skilled researchers. One of the first data set for IDS was created by DARPA which contained the data about Dos attack and Unauthorized Access [1] etc.

8.4 ML and DM Methods for Intrusion Detection

There are some of the methods for intrusion detection.

These are being discussed below:

A. Support Vector Machine (SVM)

It is a good supervised learning model, it uses associated learning algorithm so that it checks the data used for classification and regression. SVM model actually represents its example as point in space. So each sets of example can be separated according to their categories.

- **Misuse Detection and Anomaly Detection**

Misuse detection categorizes abnormal traffic (separates normal traffic from unusual traffic) in network based on clustering methods. One of the works which is mentioned in SVM classifier was used to categorize KDD 1999 Data Set into predefined Categorize (Dos, Scan, Normal). It also focused on a training subset which was founded by using Ant Colony Optimization Approach which helped out in maximize labelling of data [3].

In anomaly detection some attributes of networks are used such as source IP, destination IP, port number, type of packet and number of packets with same size. This data is collected using libcap libraries or similar libraries like in window we use wincap. These attributes of network are than formatted for SVM to learn so that it can detect anomalies in network [3].

B. Artificial Neural Network (ANN)

ANN stands for artificial neural network sometimes also called as connectionist system, the word neural is used as this method is inspired by actual brain, it is based on collection of connected units called artificial neurons. Figure 8.1 represents the connection of nodes in artificial neural network.

The above diagram represents how the nodes are interconnected, which represent networks of neurons just like in brain and each node in figure shows one artificial neuron.

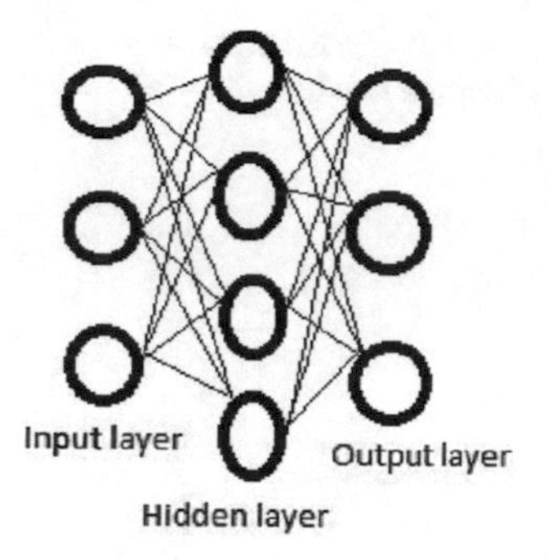

Fig. 8.1 Connection of nodes in artificial neural network

- **Misuse, Anomaly Detection**

Earlier rule-based intrusion detection was popular but it was found that it was capable only when exact characteristic of a particular attack was given but now a day's script for similar attacks varies from one attacker to another. So at this point the ANN comes in use, as it provides flexibility in analysing data of network. It can also analyse data of network in nonlinear style. It is also fast as compared to rule-based detection so it can also detect intrusion before irreversible damage is done [4].

In case of anomaly detection, it will be good in detecting such as DOS and DDOS attacks, scanning attacks as in case of DDOS attacks it can predict with help of number of packets that are coming in network and in case of scanning attacks it can check how many times a host is getting a request and from whom host is getting request as scanning attacks tries to communicate with host again and again to find open ports or some other weakness [4].

C. Decision Tree

Decision tree is a tree structure classifier. From the name itself is a structure tree having decision node and leaf node. It has two types of node, one is decision node and the other is a leaf node. It breaks down the data into smaller and smaller node. Decision tree can be used for both classification and regression however it is more popularly used for classification.

Given example: Start with the root of the tree and based on the value of the test we go to the corresponding branch and we continue doing this we come to the leaf node and the leaf node we have the value of the example; the value can be the predicted value or can be a probability. In the Fig. 8.2, P is considered as root node and P1, P2 are child node.

Mostly used methods for automatically building in decision tree are ID3 [5] and C4.5 [6]. Both algorithms decide from the set of training data which apply the concept of information of entropy.

- **Misuse Detection and Anomaly Detection**

In misuse detection the signature detection is mostly used to figure out the real signature. There are some rules defined for matching process between input signatures to actual signature. A well-known open source tool, Snort adopts signature-based

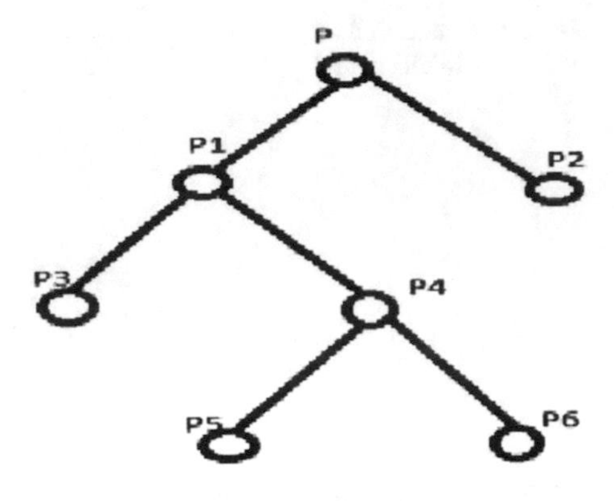

Fig. 8.2 An example of decision tree

approach. But Snort also has demerit it uses single line description which makes the matching process slow between input data and signature. Later this technique is replaced by decision tree. Kruegel and Toth [7] derived a decision tree algorithm ID3.

Other experiments were performed to determine whether the domain were malicious or not. In Anomaly detection EXPOSURE [8, 9] is a system of rules that collects large Domain Name Service and analysis to detect the domain in malicious activity. In DNS (Domain Name Service) there are some domain name created by anonymous hacker for malicious activity. The malicious activity can be detected by malware domains, the Zeus Block List, Anubis, etc.

This Techniques will work well with large amount of data sets, as large amounts of data flow across computer network so it is important to give a high performance of decision tree which will make them helpful in real-time intrusion detection.

D. Inductive Learning

Inductive learning is a process of learning by training example to achieve a given level of generalization accuracy. It generalizes from observed training examples. It generally consists of two techniques deduction and induction. Deduction deducts information that is a logical outcome of the information present in the data and continues from top to down. Inductive moves from bottom that is from particular observations to broader generalizations and theories.

Many of the researchers got a point of using the inductive learning in intrusion detection by using Repeated Incremental Pruning to produce Error Reduction (RIPPER) [10], and the algorithm quasi-optimal (AQ). An example of Ripper Rule is: if number of failed-logins is greater than or equals to 6 then this connection is a guess. RIPPER follows rules directly from the training data [11]. In Fig. 8.3.

Guess: login-failed >= 8

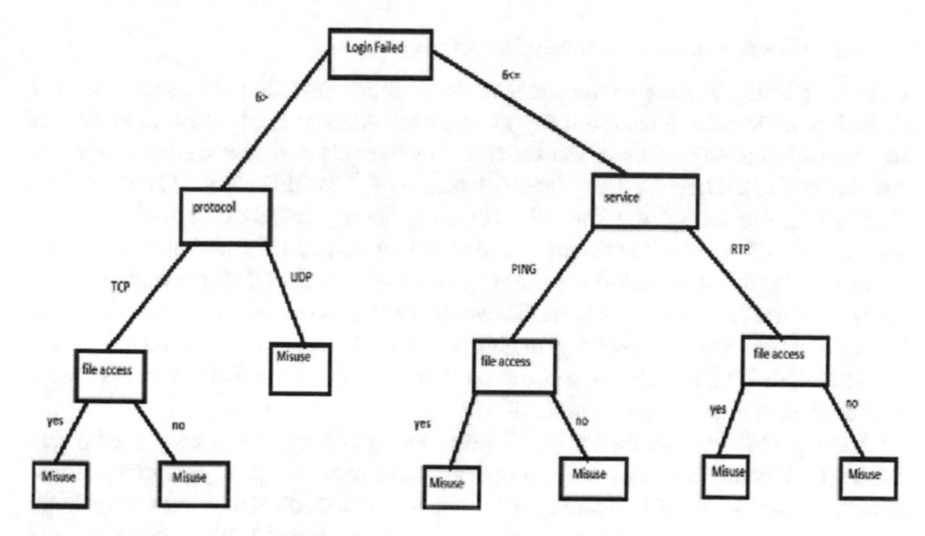

Fig. 8.3 Experiment using decision tree

- **Misuse Detection and Anomaly Detection**

A framework is created [12] in which several ML and DM techniques were applied. Statistical features were created using Sequential Pattern mining (also known as Frequent Episodes). It extracts the features from the intrusion pattern that will used in misuse detection. When the features are extracted, the RIPPER is used to generate rules. E.g. RIPPER is used to generate rules from the DARPA 1998 data set. Few weeks the data were used for training and the another few weeks the data were used for testing and finally they come to the conclusion that the test data contain 38 attacks type.

In Anomaly detection the major problem lies in discovering the boundaries between known and unknown categories [2]. To improve the performance, an anomaly generator is designed. The two approaches that are made at that time were distribution-based anomaly generation and the other one is filtered artificial anomalies. E.g. An experiment was performed on DARPA 1998 data set and come to the conclusion that 94% anomaly detection rate with 2% FAR was reached.

E. Clustering

Machine Learning Algorithm are of two kinds: Supervised Machine Learning Algorithm (SMLA) and Unsupervised Machine Learning Algorithm (UMLA). SMLA consist of input as well as output variables and an algorithm that understands the mapping function from the input variable to the output variable, so that it could predict future outputs itself with entered input. While UMLA only consist of input variables, no output variables and the algorithm are made to work on it, to observe new types of structure in data. It is unsupervised as it doesn't contain any correct answers compared to supervised learning.

- **Misuse Detection and Anomaly Detection**

Clustering is muchmore popular and used for anomaly detection compared to misuse detection, as misuse detection using clustering has fewer application as compared to anomaly detection. One of the study got recognized and was well described by its authors [13]. They used the density based model for this study. They created a Simple Log File Clustering Tool (SLCT) using density based clustering scheme. It was meant to form cluster of normal and anomaly traffic. This method could detect the new and zero day attacks that were even unseen previously. This method separated the normal traffic from anomaly traffic and the centroids of anomalous were checked by professionals in cyber security on hourly basis. This study used the KDD datasets for detections. The impressive part of this study was that it predicted about 70–80% of attacks that were not given in the KDD.

A study [14] was conducted by Blowers and Williams where they used a DB-SCAN to separate anomaly and normal network packets. They performed a test, which showed 98% of attacking packets in the detection which was very huge. Another study was conducted by Syarif, Prugel-Bennett and Wills where they used 5 clustering algorithm of k-Means, EM clustering, Distance based outlier detection algorithm, improved k-means and k-Medoids. Using these they detected the network intrusion. Distance based outlier detection had the highest performance percentage of 80% which is very high, but the false alarm rate percentage of the clustering were nearly 20% which meant to be improved in future by them. Clustering is a very useful technique for anomaly detection and many successful study and experiment have been conducted to prove it right and in future it will be even more accurate.

F. Evolutionary Computation

Evolutionary Algorithm (EA) is also known as Evolutionary Computation. EA circles around Genetic (Algorithm, Programming), Artificial Immune System, Evolution Strategies, Particle Swam Optimization and Ant Colony Optimization. Evolutionary Computation. In general, it means using techniques that are inspired from biology. The main components or the most popular amongst them are Genetic Programming (GP) and Genetic Algorithm (GA). They are based on Charles Darwin theory of Survival of the Fittest. They are judged in 3 categories mainly selection, mating and crossover. We take into consideration the chromosomes of individual. Random population is selected. Strongest individual in problem solving are usually selected for mating process and reproduction. Two of them are selected for the next process of crossover where they exchange their genetic material with each other. Finally, this may lead to mutation that is genetic alteration. This leads to a new generation of strong individuals. In Genetic Algorithm, we recognize individuals as bit string and process of mutation and crossover are easy. While in Genetic Programming as the name suggest we recognize them as programs (plus, minus, loops etc.), the process of mutation and crossover are difficult in comparison to Genetic Algorithm.

- **Misuse Detection and Anomaly Detection**

The study of Li [15] improved misuse detection using Genetic Algorithm. His study used DARPA datasets. Genetic Algorithm chromosomes contained IP Address, port number connection state etc. Mutation process and crossover are performed on population. Niching techniques were used and new sets of rules came into limelight which were added in intrusion detection. A study given by A. Abraham, Grosan and C. Martin-Vide [16] on Genetic Programming used GP to categorize attacks. They used three techniques for this (GEP) Gene expression Program, Multi Expression Program (MEP), Linear Genetic Programming (LGP). They used 1998 intrusion dataset by DARPA. Many mathematical operations like +−*/etc. were used in this study.

Anomaly Detection: GA is an efficient analysis method to find and estimate answers to search problems. Khan in his study [12] used GA for intrusion detection. Two sets of training and test were formed from KDD data sets. Surprising results were the result on test datasets were far more accurate than the results on training datasets. While Lu and Traore made a study [17] on GP using evolution approach for intrusion of novel attack and on network. They used DARPA intrusion detection datasets for this. One day for training set and one day for test set. They later on displayed only the average results of this study. Both GA and GP are efficient and used for intrusion detect. Table 8.1 discusses the methods which are used in cyber security intrusion detection. Table 8.2 shows the comparative study of the papers studied form 2002 to 2017.

Table 8.1 Methods used in cyber security intrusion detection

S. No.	Inventors name	Techniques used	Advantages	Disadvantages
1	Marvin Minsky [1951]	Artificial neural network	• Fault • Tolerance • Gradual • Corruption	• Hardware dependence • Unexplained • Behaviour of the network
2	Vladimir N. Vapnik [1963]	Support vector machine (SVM)	• Kernel trick	• Runs slow during • Testing
3	J. Ross Quinlan [1975]	Decision tree	• Intuitive knowledge • Expression • High • Accuracy	• The outcome of decisions depend majorly on • Expectation
4	–	Inductive learning	• Guess if the client is real person or anonymous	• Needed redundancy in password
5	James Macqueen [1967]	Clustering	• Easy adaptability	• Expensive
6	Lawrence J. Fogel [Early 1900]	Evolutionary computation	• Quick approximate solution	• Difficulty in encoding

Table 8.2 Comparitive study of the papers studied (2002–2017)

Year	Title of the paper	Author	Technique	Algorithm
2002	Anomaly based data mining for intrusion	Karlton Sequeira, Dr. Zakea Idris Ali	Clustering	Sequence matching algorithm
2003	Application of machine learning (ML) algorithms to KDD intrusion detection dataset within misuse detection context	Maheshkumar Sabhnani, Gursel Serpen	Neural network, clustering, decision tree	Multilayer perceptron (MLP), K means clustering algorithm, nearest cluster algorithm (NEA), Leader algorithm (LEA), C4.5 decision tree, incremental radial basis function (IRBF), hypersphere algorithm (HYP), fuzzy ARTMAP (ART)
2004	Intrusion detection systems (IDS) using decision trees and support vector Machines	Sandhya Peddabachigari, Ajith Abraham, Johnson Thomas	Decision tree, support vector machine (SVM)	ID3 algorithm, C4.5 decision tree, RIPPER algorithm
2004	Genetic algorithm for network intrusion	Wei-Li	Evolutionary computation	Genetic algorithm
2007	Genetic programming for prevention of cyberterrorism through dynamic and evolving intrusion detection	J. Hansen, P. Lowry, D. Meservy and D. McDonald	Evolutionary computation	Genetic programming (GP)
2007	Genetic programming for prevention of cyberterrorism through dynamic and evolving intrusion-detection	J. Hansen, P. Lowry, D. Meservy and D. McDonald	Evolutionary computation	Genetic programming (GP)
2007	Evolutionary design of intrusion detection programs	A Abraham, C. Grosan, C. Martin Vide	Evolutionary computation	Linear genetic programming, multi expression programming, and gene expression programming

(continued)

Table 8.2 (continued)

Year	Title of the paper	Author	Technique	Algorithm
2012	Unsupervised clustering approach for network anomaly detection	Iwan Syarif, Adam Prugel Bennett, Gary Wills	Clustering	K-means, improved k-means, k-medoids, EM clustering and distance based outlier detection algorithm
2013	Intrusion detection using support vector machine	Leena Ragha, Jayshree Jha	Support vector machine (SVM)	SVM algorithms, K-nearest neighbour algorithm
2016	A survey of data mining and machine learning methods for cyber security intrusion detection	Anna L. Buczak, ErhanGuven	Artificial neural networks, clustering, decision trees, evolutionary computation, inductive learning, support vector machine	one-class SVMs, ANNs
2017	Survey: of data mining (DM) and machine learning (ML) methods on cyber security	Rahul D. Shanbhogue, B. M. Beena	Genetic or evolutionary algorithms, artificial neural networks (ANNs), decision tree	Branch features of decision trees, Genes in genetic algorithm

8.5 Conclusion

This paper describes the overview of Machine Learning and Data Mining Techniques and Algorithms for Cyber Security in Intrusion detection. The characteristics of ML methods makes it ideal to design IDS that have higher rates of detection and low rates of false detection. Even current IDS are very effective as they will evolve with time they will become the integral and flexible part of security system. It is highly challenging for IT industries to prevent from the intrusion attacks.

8.6 Future Work

As we know that the field of intrusion detection system is comparatively new from other security systems and with vast applications so there are number of things that can be done in the field of intrusion detection like we can use ENSEMBLE learning algorithms for better accuracy we can also use soft computing with data mining techniques to increase accuracy.

References

1. Buczak, A.L., Guven, E.: A survey of data mining and machine learning methods for cyber security intrusion detection. IEEE Commun. Surv. Tutor. **18**(2), 1154–1170 (2016)
2. Fan, W., Miller, M., Stolfo, S., Lee, W., Chan, P.: Using artificial anomalies to detect unknown and known network intrusions. Knowl. Inf. Syst. **6**(5), 507–527 (2004)
3. Jha, J., Ragha, L.: Intrusion detection system using support vector machine. Int. J. Appl. Inf. Syst. (IJAIS). In: International Conference and workshop on Advanced Computing, 2013, p. 1–2. Foundation of Computer Science FCS, New York. ISSN: 2249-0868 (2013)
4. Cannady, J.: Artificial neural networks for misuse detection. School of Computer and Information Sciences Nova South-eastern University Fort Lauderadale, FL33314, pp 2–6 (1998)
5. Quinlan, R.: Induction of decision tree. Mach. Learn. **1**(1), 81–106 (1986)
6. Quinlan, R.: C4.5: Programs for machine learning. Morgan Kaufmann, San Mateo, CA (1993)
7. Kruegel, C., Toth, T.: Using decision tree to improve signature based intrusion detection. In: Proceedings of the 6th International Workshop on Recent Advances in Intrusion Detection, West Lafayette, IN, USA, pp. 173–191 (2003)
8. Bilge, L., Kirda, E., Kruegel, C., Balduzzi, M.: EXPOSURE: Finding malicious domains using passive DNS analysis. In: Presented at the 18th Annual Network and Distributed System Security Conference (2011)
9. Bilge, L., Sen, S., Balzarotti, D., kirada, E., Kruegel, C.: Exposure: A passerine DNS analysis service to detect and report malicious domains. ACM Trans. Inf. Syst. Secur. **16**(4) (2014)
10. Cohen, W.W.: Fast effective rule induction. In Proceedings of the 12th International Conference Machine Learning, pp. 115–123. Lake Tahoe, CA, USA (1995)
11. Michalski, R.: A theory and methodology of inductive learning. Mach. Learn. **1**, 83–134 (1983)
12. Lee, W., Stolfo, S., Mok, K.: A data mining framework for building intrusion detection model. In: Proceedings of the IEEE Symposium on Security and Privacy, pp.120–132 (1999)
13. Hendry, R., Yang, S.J.: Intrusion signature creation via clustering anomalities. In: Proceedings of SPIE Defence Security Symposium, International Society Optics and Photonics, pp. 69730C–69730C (2008)
14. Blowers, M., Williams, J.: Machine learning applied to cyber operations. In Network Science and Cybersecurity, pp.55–175. Springer, New York, NY (2014)
15. Li, W.: Using genetic algorithms for network intrusion detection. In: Proceedings of the United States department of energy cyber security group 2004 Training Conference, p. 1–8 (2004)
16. Khan, S.: Rule-based network intrusion detection using genetic algorithms. Int. J. Comput. Appl. **18**(8), 26–28 (2011)
17. Lu, W., Traore, I.: Detection new forms of network intrusion using genetic programming. Comput. Intell. **20**, 470–489 (2004)

Chapter 9
Performance Analysis of SVM and KNN in Breast Cancer Classification: A Survey

Ruchika Pharswan and Jitendra Singh

Abstract Breast cancer is one among the foremost widely recognized kind of cancer among feminine population in the entire world. It is still challenging task to detect and classify the cancer tumor precisely. Mammography is considered as a standout amongst the most conclusive and dependable method for proper identification and classification of the breast cancer. Here, in this paper we are proposing a system based on machine learning for classification of breast cancer (BC) along with the comparative study of two machine learning (ML) classifier. The idea is to select the region of interest (ROI) at very first from the mammograms. At that point important features has been extracted using GLCM (grey level co-occurrence matrix). Thereafter, extracted features are then utilized to train our classifiers SVM and KNN individually. The mammogram are then characterized either into benign or malignant using the trained classifier. Proposed system is implemented on standard MIAS databases. Classification performance of both classifiers are contrasted in terms of accuracy, recall, precision, specificity and F1 score. We found that SVM achieved higher accuracy of 94% than KNN with better recall and F1 score.

Keywords Breast cancer · Mammograms · Machine learning · Support vector machine · K-Nearest neighbor

9.1 Introduction

In the present 21st century, world keeps running on modern lifestyle with fast food and unhealthy habits the conceivable outcome of horrendous diseases have exaggerated exponentially. Accordingly, the early recognition of those diseases are often contrast among life and death [1]. BC is one among the foremost widely recognized kind of cancer among feminine population in the entire world. Death rate for breast cancer is beyond different kind of cancer. In 2005 BC alone caused 502,000 passing around the world [2]. Details have appeared that nearly 1.7 million new cases has been analysed in 2012 which i.e. second most basic disease in general. This speaks about 12% of all

R. Pharswan (✉) · J. Singh
SRM IST, Delhi NCR Campus, 201204 Modinagar, Ghaziabad, UP, India
e-mail: ruchi1996pharswan@gmail.com

V. E. Balas et al. (eds.), *Internet of Things and Big Data Applications*, Intelligent Systems Reference Library 180, https://doi.org/10.1007/978-3-030-39119-5_9

new cancer cases and 25% of all cancers in feminine [3]. The statistics demonstrates that 6% of ladies in India pass away because of BC. The review of 2013 shows that 230,815 women and 2019 gentlemen in the US were resolved to have BC. Assessed new BC cases and deaths by sex, US, 2017 [4].

Tumor is anomalous development of cells. Two categories of tumor are out there: (1) Benign tumor, (2) Malignant tumor inside that Benign Tumor is non-harmful, non-cancerous where as disturbing development and alarming growth of cancer is malignant. Here, AI & ML comes in picture. Over the foremost recent few decades, ML strategies are outspread in the advancement and improvement of prophetical and prognosticative models so as guiding the efficient and effective decision-making [5]. In this paper, two well know and commonly used machine learning techniques are employed to a breast cancer dataset (MIAS) in order to perform a comparative analysis. These technique are Support Vector Machine (SVM) and K-Nearest Neighbor (KNN). Their performance are analysed for their efficiency using some of the performance matrix like confusion matrix accuracy, precision, recall, specificity. The remaining part of paper is represented as follows: Sect. 9.2 explains an state-of-the-art, Sect. 9.3 our proposed methodology, Sect. 9.4 Evaluated results and interpretations, and in Sect. 9.5 conclusions and future scope.

9.2 State-of-the-Art

The foremost common and very important task in learning procedure is classification [6]. Here we have performed survey on two such ML algorithm SVM and KNN. Numerous investigations are performed on medical datasets utilizing various classifiers and feature extraction procedures. In literature, a decent measure of research on breast cancer datasets is done. A significant number of them show great classification accuracy [3].

9.2.1 Support Vector Machine (SVM)

In 2006, a completely new two-phase technique is used for segmentation of mammogram that is enforced to facilitate the automatic segmentation of small calcification which is a significant notion for breast cancer detection. The co-occurrence matrix is used in feature extraction and Support Vector Machine (SVM) is used for classifying the mammogram into benign or malignant. Their proposed system has yield the overall sensitivity of 92.8% [7].

Two years later (2008), researchers have used statistical texture features and Support Vector Machine (SVM) in collaboration in an attempt to improve the classification accuracy. They have used DDSM as a mammogram database. In order to find more accurate classifier, they have compared SVM performance with other classifiers that are LDA, NDA, PCA, ANN. However, SVM was ready to attain a higher accuracy of 82.5% [8].

Then in 2009 authors of [9], try to find out the solution for two main problems. First one is the way to detect tumors as suspicious regions with a really weak contrast and the second one is the way to extract features that can categorize tumors. The classification of a tumor will be done by using SVM classifier. And they have achieved the sensitivity of 88.75% (Table 9.1).

Later in 2011 [10], authors bestowed comparative study on SVM with various kernels to assess the detection of the breast cancer. In this examination, they tend to utilize four kernels: RBF, polynomial, Mahalanobis, and sigmoid. Accuracy of MLP is 94.63% whereas accuracy of least performed SVM kernel is 96.02%.

Well in the 2013 [11] authors presented research that considered the utilization of data mining procedures to create predictive models for recurrence of breast cancer in patients who were followed-up for quite a while. They have compared three most often used classifiers i.e., C4.5, SVM and Where they found SVM performed best with the highest accuracy, sensitivity, specificity of 95.7%, 97.1%, 94.5% respectively. Then in 2016, authors of [12], have proposed a system for classifying breast

Table 9.1 Conclusion of work done with SVM in last few years

Year	Investigators	Database	Features	Classifiers	Accuracy (%)
2006	Selvil et al. [7]	NA	Ridgelet transformation and co-occurrence matrix	SVM	92.8
2008	Abdalla et al. [8]	DDSM	Statistical feature and co-occurrence matrix	Comparative study, SVM with higher accuracy	82.5
2009	Rejani et al. [9]	Mini-MIAS	Morphological features	SVM	88.75
2011	Hussain et al. [10]	WBCD	Genetic algorithm	SVM kernel	96.02
2013	LG et al. [11]	ICBC	NA	Comparative study, SVM with higher accuracy	95.7
2016	Kanchanamani et al. [12]	MIAS	Statistical feature and GLCM	Comparative study, SVM with higher accuracy	92.5
2017	Ancy et al. [13]	MIAS	GLCM	Linear SVM	81
2017	Jothilakshmi et al. [14]	Mini-MIAS	GLCM	Multi-SVM	94
2018	Karthiga et al. [15]	Breakhis	Coifelt wavelet	Linear-SVM	93.3

cancer, wherein they used Shearlet Transform and have compared various classifier's performances in order to choose the best one. On the basis of their comparison experiment, they concluded SVM classifier produces the highest accuracy (92.5%) at the point when contrasted with different classifiers results.

Another study in 2017 [13], proposed an approach for classifying mammograms where they used Gray Level Co-occurrence Matrix (GLCM) and Support Vector Machine (SVM). They achieved accuracy of 81% and sensitivity of 73%. Within same year, recently somewhere similar approach is followed by other authors [14], with an objective of increasing the accuracy of classification using GLCM, where they have selected 13 features and finally mammogram is classified into benign or malignant on the basis on multi-SVM prediction. And approach has given accuracy of 94%.

Recently 2018, Karthiga et al. has proposed another study where first they have performed pre-processing via image enhancement then have done segmentation using K-means clustering, later features are extracted by Coifelt wavelet, finally tumor is classified into benign or malignant using linear SVM and achieved accuracy of 93.3% [15].

9.2.2 K-Nearest Neighbor (KNN)

As early as 1996, a strategy is projected for the detection of groups of micro calcifications. The tactic first segments the image into suspected areas utilizing morphological filters. After that, for undertaking the classification of the region as normal or MC they have taken K-NN classifier into account with two different distances. One is Euclidean distance measure and another is locally optimum distance measures. The best classification outcome 0.94 is delivered utilizing the locally optimum distance measure at $K = 2$ [16].

Later in 2013, authors [17] did a research on KNN algorithm with various distance measures and also classification rules to improve the performance of breast cancer diagnosis. After all things considered, they concluded these two have given best outcome: at $k = 1$, Euclidean distance −98; 70%, Manhattan distance −98; 48%. After a year, Ohmshankar et al. [18], proposed their research for analysis and early discovery of breast cancer by utilizing mammogram pictures. Thirteen haralick features are extracted by employing Gray Level Co-occurrence Matrix (GLCM). For classification they have employed robust K-Nearest Neighbour (K-NN) and have acquired the accuracy of 92%.

Within same year, two-phase classification strategy for the characterization of microcalcification was given by authors [19] with a comparative scrutiny on (DCT) and (DWT). It is seen from their experiments that DCT based mostly classification system gives agreeable outcome over DWT with most extreme accuracy of 93.86% whereas DWT based system accomplishes just accuracy of 90.43%.

As year passes more researches are taken into account in order to improve the accuracy of classification of the breast cancer, similarly one more study presented

Table 9.2 Conclusion of work done with K-NN in last few years

Year	Investigators	Database	Classifiers	Distance measure used	Value of "k"	Accuracy (%)
1996	Hojijatoleslami et al. [16]	MIAS	KNN	Locally optimum distance	K = 2	94
2013	Medjahed et al. [17]	WBCD	KNN	Euclidean Distance	K = 1	70
2014	Ohmshankar et al. [18]	MIAS	KNN	Euclidean distance	K = 1	92
2014	Bala et al. [19]	MIAS	KNN	NA	NA	91
2017	Dong et al. [20]	MIAS	Improved KNN	Euclidean distance	K = 3	95.76
2018	Amrane et al. [21]	WBCD	Comparison study, KNN have higher accuracy	Euclidean distance	K = 3	97.51

by authors in 2017 [20] where they employed improved KNN for classification and Dual Contourlet Transform (Dual-CT) is used for feature extraction. Where they found improved KNN perform better with 95.76% of accuracy.

And recently in 2018, authors [21] have made a comparative study on two classifiers which are: Naïve Bayesian (NB) and k-Nearest Neighbors (KNN), their research evidence indicates that KNN accomplished a higher productivity of 97.51%, nonetheless, even NB has an accuracy of 96.19% (Table 9.2).

9.3 Methodology

This part depicts about the proposed system. It begins with selecting the Region of Interest (ROI) and from these extracted ROI GLCM assigned texture feature are derived to make feature vector illustration for every mammogram picture. These calculated feature vectors are then given to classifier as an input, to characterize input mammogram into Benign and malignant [22]. Here for classification, two classifiers have been attempted and tried. SVM and KNN are the classifiers utilized. MATLAB is utilized for this analysis study [3]. We make use of Mammographic Image Analysis Society (MIAS) database. We have utilized total of 150 mammogram out of which 100 are used as training dataset and rest as testing dataset. For the sake of symmetry our training dataset is comprise of 50 benign mammograms and 50 malignant mammograms. And likewise in our testing dataset we have two halves of 25 one half is for benign mammograms and another for malignant. Region of Interest is then selected from each mammogram. This is done by using the information given in the MIAS database as there for x and y coordinate of the centre of abnormality is given also with the radius (r) to which it extents. This simplifies the process selecting the region of interest. Feature extraction and selection is the method of choosing relevant features. There are 14 statistical features were proposed by Haralick et al. [23] out of which we have used seven only. We have used GLCM for feature extraction [4]. We have utilized following 7 features: autocorrelation, contrast, variance, correlation, sum entropy, information measure of correlation and inverse different moment. We carried out a comparative study on two widely used machine learning classifiers, one is Support Vector Machine (SVM) another is K-Nearest Neighbour (K-NN). We have tested both classifiers individually and then have compared their performance against their evaluated confusion matrix, accuracy, recall, specificity, precision and F1 score. We have utilized confusion matrix, accuracy, recall, precision, Specificity, F1 score as a performance evaluation parameters.

9.4 Results and Interpretation

The comparative results on two classifiers SVM and KNN is shown in Figs. 9.1, 9.2 and Table 9.3.

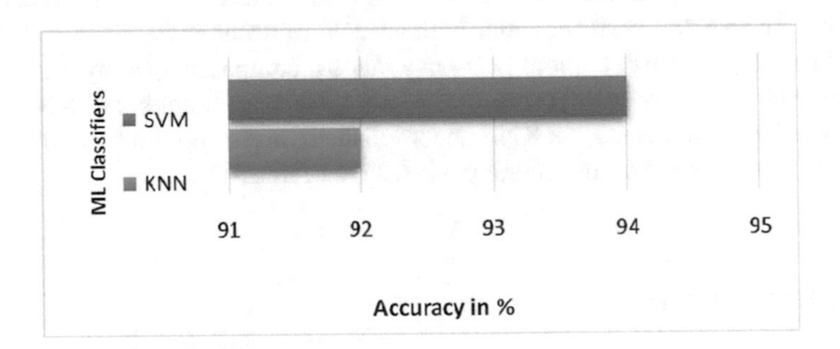

Fig. 9.1 Comparison of accuracy of SVM and KNN

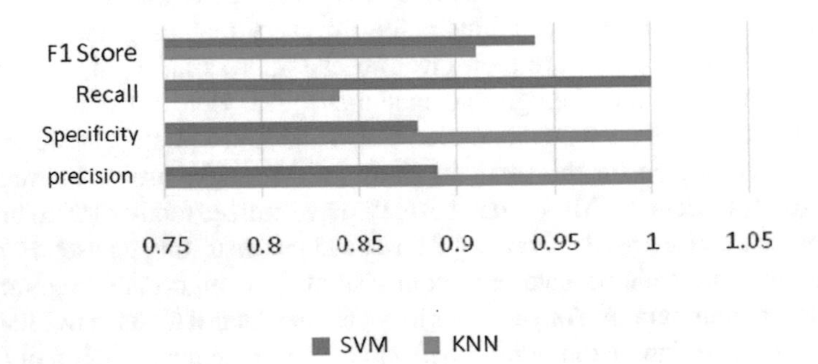

Fig. 9.2 Comparison of performances of SVM and KNN

Table 9.3 Comparison of confusion matrix of SVM and KNN

Classifier	True Positive (TP)	True Negative (TN)	False Positive (FP)	False Negative (FN)
K-Nearest Neighbour (KNN)	21	25	0	4
Support vector machine (SVM)	25	22	3	0

9.5 Conclusion and Future Scope

In this paper, we represent our comparative research on powerful and widely used machine learning algorithms SVM and KNN that are used for classifying breast cancer into benign or malignant. Our primary objective is to discover best one among these two on the basis of their performance. We found that, SVM achieved higher accuracy of 94%, better F1 score and recall than KNN. However, KNN wasn't far behind, it also has shown quite good performance in terms of specificity and precision. Since cancer is one of the widespread and death causing disease among women, accurate detection and correct classification is extremely crucial for its better treatment thus the classification accuracy is the biggest concern. As a result, there is a constant requirement for improving the accuracy. On the basis of our comparative analysis results we will make use of SVM classifier for classifying the cancer into benign or malignant and we will also try to reduce the number of features by selecting optimal features and hence can reduce the training overhead too.

References

1. Mangale, S., Thombre, V.: A Survey on efficient image retrieval through user clicks and query semantic signature. Int. J. Innov. Res. Comput. Commun. Eng. (An ISO Certified Organization) **3297**(5), 21167–21175 (2007)
2. Johra, F.T., Shuvo, M.M.H.: Detection of breast cancer from histopathology image and classifying benign and malignant state using fuzzy logic. In: 2016 3rd International Conference on Electrical Engineering and Information Communication Technology (ICEEICT) 2016, vol. 1, pp. 1–5 (2017)
3. Saraswathi, D., Srinivasan, E.: Performance analysis of mammogram CAD system using SVM and KNN classifier. Proc. Int. Conf. Inven. Syst. Control. ICISC **2017**, 1–5 (2017)
4. Ghongade, R.D., Wakde, D.G.: Computer-aided diagnosis system for breast cancer using RF classifier, pp. 1068–1072 (2017)
5. Noor, M.M., Narwal, V.: Machine learning approaches in cancer detection and diagnosis : Mini review machine learning approaches in cancer detection and diagnosis : Mini review (2017)
6. Kourou, K., Exarchos, T.P., Exarchos, K.P., Karamouzis, M.V, Fotiadis, D.I.: Machine learning applications in cancer prognosis and prediction. **13**, 8–17 (2015)
7. Selvi, S.T., Malmathanraj, R.: Segmentation and SVM classification of mammograms. In: proceedings of the IEEE International Conference on Industrial Technology, pp. 905–910 (2006)

8. Science, C., Systems, I., Abdalla, A.M.M., Deris, S., Zaki, N., Ghoneim, D.M.: Breast cancer detection based, pp. 728–730 (1841)
9. Rejani, Y.I.A., Selvi, S.T.: Early detection of breast cancer using SVM classifier technique. **1**(3), 127–130 (2009)
10. Hussain, M., Wajid, S.K., Elzaart, A., Berbar, M.: A comparison of SVM kernel functions for breast cancer detection. In: Proceedings of 2011 8th International Conference Computer Graphics, Imaging and Visualization CGIV 2011, pp. 145–150 (2011)
11. Lg, A., At, E.: Using three machine learning techniques for predicting breast cancer recurrence. J. Heal. Med. Inform. **04**(02), 2–4 (2013)
12. Kanchanamani, M., Perumal, V.: Performance evaluation and comparative analysis of various machine learning techniques for diagnosis of breast cancer. Biomed. Res. **27**(3), 623–631 (2016)
13. Ancy, C.A., Nair, L.S.: An efficient CAD for detection of tumour in mammograms using SVM. In: Proceedings of 2017 IEEE International Conference on Communication and Signal Processing ICCSP 2017, vol. 2018, pp. 1431–1435 (2018)
14. Jothilakshmi, G.R., Raaza, A.: Effective detection of mass abnormalities and its classification using multi-SVM classifier with digital mammogram images. In: International Conference on Computer, Communication and Signal Processing Special Focus IoT, ICCCSP 2017, pp. 1–6 (2017)
15. Karthiga, R., Narasimhan, K.: Automated diagnosis of breast cancer using wavelet based entropy. In: Second International Conference on Electronics, Communication and Aerospace Technology (ICECA), pp. 274–279 (2018)
16. Classifier, K.N., Hojijatoleslami, S.A., Ecittler, J.: Detection of clusters of microcalcificatioin using a k-nearest neighbour classifier, pp. 1–6, 1996
17. AhmedMedjahed, S., Ait Saadi, T., Benyettou, A.: Breast cancer diagnosis by using k-nearest neighbor with different distances and classification rules. Int. J. Comput. Appl. **62**(1), 1–5 (2013)
18. Visagie, S.: Classification system. January, 409–413 (2019)
19. Bala, B.K., Audithan, S.: Wavelet and curvelet analysis for the classification of microcalcifiaction using mammogram images. In: International Conference on Current Trends in Engineering and Technology ICCTET 2014, pp. 517–521 (2014)
20. Dong, M., Wang, Z., Dong, C., Mu, X., Ma, Y.: Classification of region of interest in mammograms using dual contourlet transform and improved KNN. J. Sensors, **2017** (2017)
21. Amrane, M.: Breast cancer classification using machine learning. In: *2018* Electric Electronics, Computer Science, Biomedical Engineering Meeting. pp. 1–4
22. Sonar, P., Bhosle, U., Choudhury, C.: Mammography classification using modified hybrid SVM-KNN. In: Proceedings of IEEE International Conference on Signal Processing and Communication ICSPC 2017, vol. 2018–Jan(July), 305–311 (2018)
23. Kumar, V., Gupta, P.: Importance of statistical measures in digital image processing. Int. J. Emerg. Technol. Adv. Eng. **2**(8), 56–62 (2012)

Chapter 10
Review of MAC Protocols for Energy Harvesting Wireless Sensor Network (EH-WSN)

Vivek Kumar Verma and Vinod Kumar

Abstract Traditional wireless sensor network (WSN) consist of large number of small and inexpensive battery-powered wireless sensor nodes. It is used in an unattended manner to monitor various environmental, structural and healthcare activities. Traditional WSN has limited lifetime because it is battery powered, so to enhance (1) Energy Efficiency as well as (2) Quality of service Ambient Energy in the form of solar, thermal, RF Energy, etc. are used to power wireless sensor nodes as well as to recharge the battery. But Due to uncertainty in harvested energy there arises a need for redesigning of MAC protocols and routing protocols for the Energy Harvested wireless sensor network (EH-WSN). In this paper survey of various MAC protocols for Energy harvested wireless sensor network is presented and it is based on different parameters such as Energy Efficiency, end-to-end delay, Quality of service, scalability, etc. Survey is presented in the order of proposed year so one can easily find various gaps in the subsequent studies and get the idea of future research.

Keywords Energy harvesting · MAC protocols · Sensor first section

10.1 Introduction

Today WSN is required everywhere to observe various environmental and physical parameters over a large area. Application includes detection of natural calamities such as (1) forest fire detection, (2) early notification for flood, (3) water pollution, (4) Air pollution, (5) Body area Network (BAN). All the above applications require a sensor node which can vary from very few to thousand in numbers. So one requires to develop a network which can be of (1) Mesh Type (2) Star Topology (3) Hybrid Topology etc. in which role of the sensor can be either (a) Source Node (to sense data) (b) Relay node (to relay the data of source node) and (c) Sink Node (Gateway).

V. K. Verma (✉) · V. Kumar
Department of Electronics and Communication Engineering, SRM Institute of Science and Technology, Ghaziabad, India
e-mail: vivek.verma@abes.ac.in

V. Kumar
e-mail: vinodkumar.viet@gmail.com

V. E. Balas et al. (eds.), *Internet of Things and Big Data Applications*, Intelligent Systems Reference Library 180, https://doi.org/10.1007/978-3-030-39119-5_10

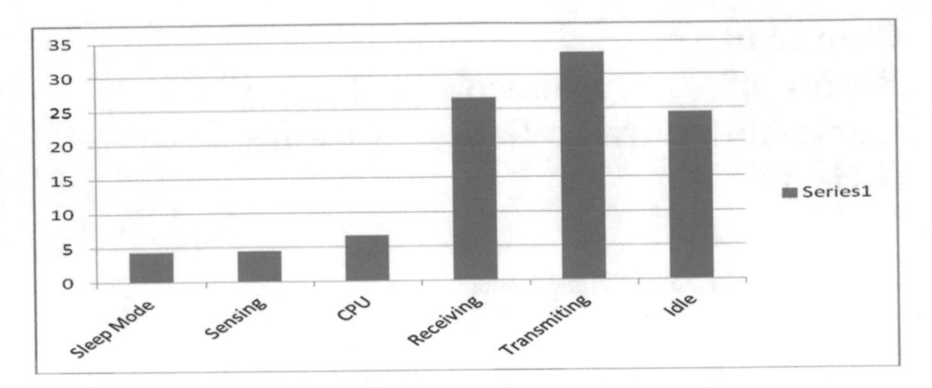

Fig. 10.1 Energy consumption profile of sensor node in term of percentage

Since Range of Individual sensor is limited so communication takes place in the multi-hop method.

Sensor nodes have basically five components: (1) Sensors (2) Processor (3) Transceiver (4) Power source (5) Energy Harvester (optional). In most of the application, Sensor nodes operate in hard to reach environment so it's very difficult to replace the battery when it is completely drained out and among all the components of sensor node major power source of power consumption are transmitter and receiver section which is depicted in Fig. 10.1 [1].

EH-WSN increases the lifetime as well as throughput as compare to traditional WSN. It includes harvested energy in the form of Thermal, RF Energy, and Solar Energy etc. With the careful design of MAC protocol it can also have infinite life if it operates in ENO-MAX condition. In this condition the Consumption rate should be less than or equal to harvesting rate. After the Introduction paper is organize as follows in Sect. 10.2 classification of MAC protocols for EH-WSN is presented. Section 10.3 deals with the previous survey conducted in this area and at the last comparison between various MAC protocol is presented with respect to some parameters.

10.2 Classification of MAC Protocols

Mac protocols classified as either synchronous or asynchronous protocols since harvesting rate is variable and depends upon environmental conditions so synchronous protocol is not preferred in EH-WSN [2]. Asynchronous protocols are further classified as (1) Sender initiated (2) Receiver initiated and (3) Sink initiated. Receiver initiated protocols are more preferred due to negligible idle listening, reduce overhearing nd it can recover lost data due to collision. Classification of MAC protocols for Energy Harvested Wireless Sensor Network is shown in Fig. 10.2. When MAC Protocol designed some important parameters that are needed to be considered are [2]: (1) Throughput (2) Packet loss (3) Latency (4) Scalability (5) Hop Count (6)

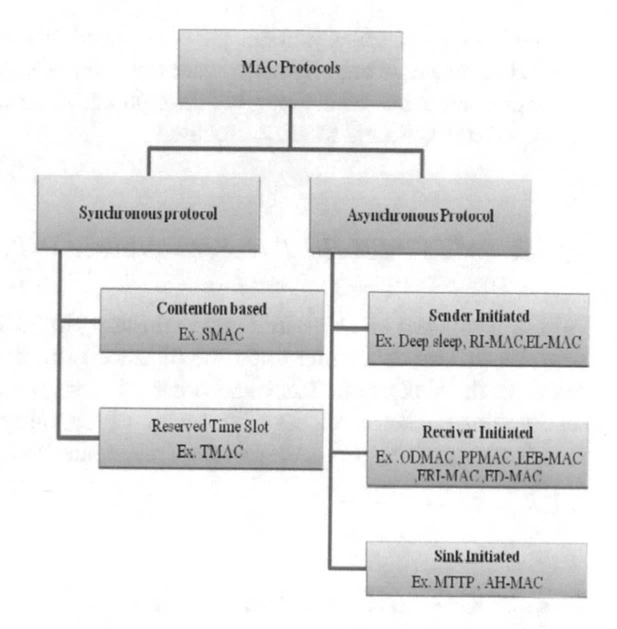

Fig. 10.2 A classification of MAC protocols

Data Aggression (7) Load Balancing (8) Error Correction and detection. Design of MAC protocol for EH-WSN also depends upon application such as (1) Continuous monitoring (2) Event Driven monitoring (3) Fixed interval monitoring.

10.3 Related Work

Several authors presented a survey on MAC protocols on the basis of various parameters such as Iannello et al. [3] Focus on analysis of conventional MAC protocols such as TDMA, Frame ALOHA for EHWSN. In This paper time efficiency and delivery efficiency is tested using Markov Models. Del et al. [4] Survey focus only on Asynchronous MAC protocols such as ODMA, XMAC, EA-MAC, EH-MAC, P-MAC, SMAC TAMC and comparison between ODMAC and X-MAC, EH-MAC compared with WSF-MAC based on Related parameters.

10.4 MAC Protocols for EH-WSN

10.4.1 PP-MAC (Probabilistic Polling MAC)

It is single-hop protocol [5] can be slotted or unslotted CSMA protocol and in this instead of sending ID the contention probability Pc is transmitted. when node wants to send data it generates number between 0 and 1 if it is less than Pc then it will allowed to transmit data otherwise not and this protocol is simulated by Qualnet Simulator and achieve considerable throughput, fairness flexibility or scalability as compared to traditional MAC protocols.

10.4.2 MTTP (Multi Tier Probabilistic Polling)

Fujii [4] Proposed this sink initiated multi-hop protocol in this all sensor nodes are arranged in Tire depending upon the distance from the sink the sensor and nodes nearer to the sink are in Tier1 and other are assigned tier no. 255. whenever sink send polling packet it consist of contention Probability Pc along with tier number. Disadvantage of this protocol is that overhead increases with the increase in number of tier.

10.4.3 OD-MAC (On Demand MAC)

This protocol is a multi-hop receiver-initiated protocol proposed by Agoutis et al. [6] in this protocol receiver ask for the transmission of data by the transmitter by sending a small beacon. This protocol supports individual duty cycle of the node which is having a different level of energy and also operates in ENO max state. This protocol also reduces delay by using an opportunistic forwarding scheme [7] and in this protocol idle listening shifted from receiver to transmitter. Working of a protocol is shown in Fig. 10.3. ODMAC is implemented in OMNET simulator and the parameters that are considered in this simulation are: (1) Performance versus Energy conservation (2) Achieving ENO state (3) Energy Availability versus Performance (5) Load balancing.

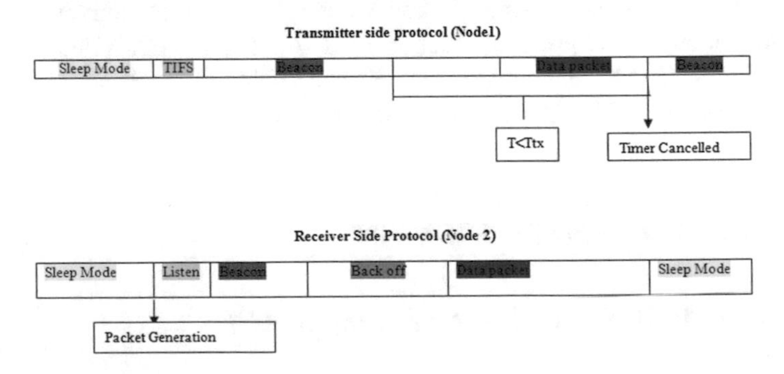

Fig. 10.3 Working of ODMAC protocol

10.4.4 EH-MAC

It is a receiver-initiated multi-hop Asynchronous Protocol proposed by Fafoutis et al. [8] This is a probabilistic polling method to reduce collision of the packet. In this protocol to reduce the interference, it dynamically adjusts the polling packet according to energy harvested. The author used QUALNET Simulator to test the performance of protocol and compare throughput with X-MAC and RI-MAC it is concluded that EH-MAC outperforms RI-MAC by 27%.

10.4.5 QAEE-MAC

QAEE-MAC is the receiver-initiated Asynchronous protocol developed by Kim et al. [9]. In this receiver wake up periodically and receive beacon send by the sender and based on senders priority receiver will broadcast Beacon Frame having the sender's address this protocol adjusts its duty cycle based on energy state.

10.4.6 ML-MAC

In this protocol, nodes are arranged in layers which results in less traffic or the chance of collision due to this life of sensor network increases [10]. In this listen period of the sensor is also reduces which further results in lesser power consumption as compared to S-MAC but in this protocol data packets arc buffered which results in larger delay or throughput. Simulation is performed in MATLAB and compared with respect to S-MAC which outperforms the S-MAC by 54% in energy efficiency.

10.4.7 HE-MAC (Harvested Energy MAC)

This protocol proposed by Lee et al. [11] Outperforms F-ALOHA which uses fixed-size frame but in this protocol, the variable size frame is used by considering the residual energy of the device. In this throughput of F-ALOHA MAC is compared with HE-MAC and with increasing energy harvesting probabilities it performs the F-ALOHA in terms of throughput.

10.4.8 ERI-MAC (Energy Harvested Receiver Initiated MAC Protocol)

This protocol is also the Receiver initiated Asynchronous Protocol developed by Nguyen et al. [12] based on energy harvesting and this protocol provides high throughput and energy efficiency it adjusts its harvesting rate depends upon the environment and also used queuing approach and simulation is performed with NS-2. In this whenever the ratio of consumed energy to harvested energy is less than one then it will operate in active state when it is greater than one it goes to sleep state or recharge state.

10.4.9 LEB-MAC (Load and Energy Balancing MAC Protocol)

This protocol is receiver-initiated asynchronous protocol proposed by Liu et al. [13] and helps in balancing the load based on energy harvesting characteristics of the node by proper scheduling of the data based on receiver energy state. In this author compared LEB-MAC [13] with RI-MAC, PW-MAC and simulation results show that it outperforms the above protocols in terms of the end to end latency, duty cycle, packet delivery ratio and collision ratio and simulation performed with QUALNET simulator.

10.4.10 Deep Sleep (Enhance Version of IEEE 802.11)

This protocol increases the energy efficiency of IEEE 802.11 by using a deep sleep algorithm. A deep sleep algorithm [14] consists of two types of an algorithm (1) Energy-aware algorithm (2) controlled Access algorithm. Energy-aware algorithm increases the sleep time of devices having high channel access priority but low energy. Controlled Access algorithm reduces the contention level by reducing the number of active devices contending for channel access. The proposed system based on single-hop rather than multi-hop.

10.4.11 EDF-MAC (Earliest Deadline First Polling MAC Protocol)

The requirement of energy varies from node to node so harvesting rate is also varied so the proposed protocol [15] regulated the transmission pattern of each node and reach to that condition where optimal trade off between fairness and channel utilization is achieved. In this proposed model single-hop network is considered where the sink node polled to each sensor node. Here sensor node calculates it's harvesting rate and then transmit it along with next predicted wakeup time to the sink node. The simulation is performed with Qualnet 5 Simulator and performance compared with HEAP-PP and HEAP-CSMA with respect to fairness versus link error probability and channel utilization versus link error probability parameter.

10.4.12 WURMAC (Wake up Radio-Based MAC Protocol)

As earlier protocol it is not based on the duty cycle approach Oller et al. [16] proposed this protocol which is based on the wake-up radio concept instead of the traditional duty cycle concept. In this secondary radio which consumes microampere current is used to wake up the main trans-receiver and microcontroller unit. This protocol simulated with OMNET++ simulator and compared with an existing protocol like BMAC, XMAC, RIMAC, and IEEE 802.15.4.

10.4.13 QPPD-MAC

This [9] protocol based on the priority level of data and transmits higher priority level of data with minimum end to end delay while considering the energy level of the receiver and this protocol is verified using actual solar readings. It is observed that end to end delivery delay is reduced by 54% and does not depend on the level of irradiance and it also give high throughput, good packet delivery ratio as compared to QAEE-MAC [10] and ERI-MAC [12].

By comparative study it comes to now most of the work done far is in the area of receiver-initiated protocol because it decreases idle listening time and hence results in less energy consumption and it is also concluded from the study that most of the protocol based on duty cycling approach but some studies reveal the use of newer technologies which is based on wake-up radio concept.

10.5 Conclusion

Energy harvested wireless sensor network have a longer life than traditional battery powered wireless sensor network and it also give high throughput as compare to later one but since harvested energy is un predictable in nature so one have to carefully design MAC protocol to achieve desirable performance. This review paper discuss all the MAC protocol with their advantage and disadvantage and it is also concluded

Table 10.1 Comparative analysis of MAC protocols for EH-WSN

Protocol	Mechanism used	Features
PP-MAC	CSMA/CA	Single hop, high throughput, scalability and fairness and moderate end to delay
MTPP	CSMA/CA	Multi hop, high throughput, scalability but overhead increases with number of tier
ODMAC	Receiver initiated	Single hop, moderate throughput and fairness but low latency and scalability
EH-MAC	Receiver initiated	Single hop, high throughput, scalability, fairness and latency
QAEE-MAC	Receiver initiated	Single hop, high throughput, provide good quality of service and improves end to end delay
ML-MAC	Contention based	Single Hop, advantage-less number of collision, disadvantage-large delay due to data buffering
HE-MAC	Based on F-ALOHA	Single hop, high throughput
ERI-MAC	Receiver initiated	Single hop, high throughput and energy efficiency
LEB MAC	Receiver initiated	Multi hop, moderate throughput and fairness but having low latency and scalability
Deep Sleep	IEEE 802.11	Single hop, useful for machine to machine communication but suffer from overhearing and idle listening
EDF-MAC	Sink initiated	Single hop, achieved better channel utilization and fairness
WURMAC	Wake up radio concept	High energy efficiency and low latency
QPPD MAC	Receiver initiated	Single hop, send high priority packet with minimum end to delay, high throughput

that no protocol discussed so far included the concept of mobility and security feature so one can go further to do research in this area. Table 10.1 represent summary of Comparison of various MAC protocol.

References

1. Engmann, F., Katsriku, F.A., Abdulai, J.-D., Adu-Manu, K.S., Banaseka, F.K.: Prolonging the lifetime of wireless sensor networks: A review of current techniques. Wirel. Commun. Mob. Comput. **2018**, 1–23 (2018)
2. Kaur, P., Sohi, B.S., Singh, P.: Recent advances in MAC protocols for the energy harvesting based WSN: A comprehensive review. Wirel. Pers. Commun. **104**(1), 423–440 (2019)
3. Iannello, F., Simeone, O., Spagnolini, U.: Medium access control protocols for wireless sensor networks with energy harvesting, pp. 1–26
4. Eu, Z.A., Tan, H.P., Seah, W.K.G.: Design and performance analysis of MAC schemes for wireless sensor networks powered by ambient energy harvesting. Ad Hoc Netw. **9**(3), 300–323. LNCS. Homepage, http://www.springer.com/lncs (2011). Accessed 21 Nov 2016

5. Del Testa, D., Marin, G., Peretti, G.: Comparison of MAC techniques for energy harvesting wireless sensor networks
6. Fujii, C., Seah, W.K.G.: Multi-tier probabilistic polling in wireless sensor networks powered by energy harvesting. In: Proceedings of the 2011 7th International Conference on Intelligent Sensors, Sensor Networks and Information Processing ISSNIP 2011, pp. 383–388. LNCS. Homepage, http://www.springer.com/lncs (2011). Accessed 21 Nov 2016
7. Fafoutis, X., Dragoni, N.: Odmac. August 2014, 49 (2011)
8. Eu, Z.A., Tan, H.: Probabilistic polling for multi-hop energy harvesting wireless sensor networks, pp. 271–275. LNCS. Homepage, http://www.springer.com/lncs (2012). Accessed 21 Nov 2016
9. Kim, S.C., Jeon, J.H., Park, H.J.: QoS aware energy-efficient (QAEE) MAC protocol for energy harvesting wireless sensor networks. In: Lecture Notes in Computer Science (including subseries Lecture Notes in Artificial Intelligence and Lecture Notes in Bioinformatics), vol. 7425, pp. 41–48. LNCS (2012)
10. Jha, M.K., Pandey, A.K., Pal, D., Mohan, A.: An energy-efficient multi-layer MAC (ML-MAC) protocol for wireless sensor networks. AEU Int. J. Electron. Commun. **65**(3), 209–216 (2011)
11. Lee, H.-K., Lee, M., Lee, T.-J.: Harvested energy-adaptive MAC protocol for energy harvesting IOT networks, pp. 51–58 (2015)
12. Nguyen, K., Nguyen, V.H., Le, D.D., Ji, Y., Duong, D.A., Yamada, S.: ERI-MAC: An energy-harvested receiver-initiated MAC protocol for wireless sensor networks. Int. J. Distrib. Sens. Netw. **2014**(May) (2014)
13. Liu, H.I., He, W.J., Seah, W.K.G.: LEB-MAC: Load and energy balancing MAC protocol for energy harvesting powered wireless sensor networks. In: Proceedings of the International Conference on Parallel and Distributed Systems (ICPADS), vol. 2015, pp. 584–591 (2014)
14. Lin, H.: DeepSleep IEEE 802.11 enhancement for energy-harvesting machine-to-machine communications, pp. 5231–5236 (2012)
15. Jin, Y., Tan, H.P.: Optimal performance trade-offs in MAC for wireless sensor networks powered by heterogeneous ambient energy harvesting. In: 2014 IFIP Networking Conference IFIP Networks (2014)
16. Oller, J., Demirkol, I., Casademont, J., Paradells, J., Gamm, G.U., Reindl, L.: Has time come to switch from duty-cycled MAC protocols to wake-up radio for wireless sensor networks? IEEE/ACM Trans. Netw. **24**(2), 674–687 (2016)

Chapter 11
Design and Analysis of Wideband Antenna for Biomedical Purposes

Ankit Kumar Chaubey, Amit and Satya Sai Srikant

Abstract A wideband rectangular formed micro strip patch antenna has been intended for biomedical purpose covering the 2.45 GHz recurrence band. Two parallel openings are joined to irritate the surface current way, presenting neighbourhood inductive impact that is in charge of the excitation of the second resounding mode. The length of the middle arm can be cut to tune the recurrence of the second thunderous mode without influencing the major resounding mode. An exhaustive parametric investigation has been done to comprehend the impacts of different dimensional parameters and to streamline the execution of the antenna. A substrate of low dielectric consistent is chosen to get a minimal transmitting structure that meets the requesting data transfer capacity determination. The reflection coefficient at the contribution of the upgraded rectangular formed micro strip patch antenna is underneath −10 dB over the whole recurrence band. The estimation results are in fantastic concurrence with the HFSS 15.v re-enactment results.

Keywords Biomedical purpose · Wideband · Microstrip patch · Prove feed · Health monitoring

11.1 Introduction

In as shown by infection recurrence bits of knowledge in 2018, 2.25 million people were resolved to have dangerous development worldwide and 0.8 million people passed on from threatening development, in India. There are various components to impact this, for instance, lifestyle, sustenance's, contamination, stress and dubious things. The most basic methodology before recovering is b assurance. Prior finding

A. K. Chaubey · Amit (✉) · S. S. Srikant
SRM Institute of Science and Technology, Modinagar, Ghaziabad, UP, India
e-mail: amitsinghrathi.81@gmail.com

A. K. Chaubey
e-mail: ankitkumarchaubey@gmail.com

S. S. Srikant
e-mail: satya.srikant@gmail.com

V. E. Balas et al. (eds.), *Internet of Things and Big Data Applications*, Intelligent Systems Reference Library 180, https://doi.org/10.1007/978-3-030-39119-5_11

has critical centrality is inspiration for returning back to life for certain suffering. Various techniques are there to recognize ailment, for instance, X-shaft, ultrasound, tomography and X-beam. Nevertheless, they have some negative and undesired consequences. Especially in case of progressively young patients, such type of systems isn't supported because of the use of ionized radiation.

In the new overall development, heaps of new manifestations had been made in biomedical telemetry. One of them is implantable helpful contraption (IMDs). IMDs have the competency to talk about remotely with an outside contraption, and have gotten goliath thought for achieving persistent and set away physiological data in biomedical telemetry. It is moreover skilled in different applications, for instance, temperature screens, blood-glucose sensor, cochlear and retina embeds, similarly as pacemakers and cardioverter defibrillators. An implantable antenna is an essential part towards radio repeat (RF)-associated IMDs. In any case, it goes with various hindrances due to the uninformed rates, kept range between the implant and outside screen, similarly as difficult to use for home watching. Prosperity issues should be seen as when an antenna is installed inside the human body insinuating IEEE C95.1 regulated definition. Thusly, to beat these repressions, some investigation is by and by organized towards implantable antenna for biomedical telemetry.

Rising usages of the microstrip fix antenna consolidate their necessities in biomedical division too. These creating applications incorporate remote prosperity checking, glucose watching, self-checking, stomach related watching, etc. In this way, with the extending need, the circumstance gets changed to the use of fix antenna in biomedical zone. On the other hand, with the brisk addition of diseases, people are right now twisting up progressively careful for their prosperity. There are various afflictions, for instance, diabetes, illness, sadness, thyroid, tuberculosis and HIV/Aides in which the human body out of the blue put on or get progressively fit. In these cases, a patient or an individual went to the pro or ace for standard enrolment. With the usage of fix antenna in a checking machine, a patient or an individual need not to go to the therapeutic facility or rec focus, and his/her weight can be successfully seen at home as well.

To design an antenna for biomedical purposes, one essential idea to be secured is that the proportion of the antenna should be close to nothing, and it performs extraordinary helpfulness. Microstrip fix antennas offer perfect inclinations, for instance, decreased size, negligible exertion and straightforward creation. In this way, these antennas are the strong plausibility to be used for biomedical purposes. In this paper, a rectangular shaped meandered opening microstrip fix antenna resonating at 2.45 GHz repeat is recommend. In light of, a single repeat model can be gotten using a rectangular framed structure. The recommend antenna works in the ISM (current Intelligent and Restorative) band. In view of its wide information transmission characteristics, this antenna can in like manner be used in checking machine with the ultimate objective of remote prosperity watching. It construes that if the repeat gets moved from its one of a kind regards, it will be inside that repeat go because of antenna's wideband features. Thusly, this antenna can be used for remote prosperity watching similarly as weight checking. The effect of dielectric properties and insert significance are in like manner discussed (Fig. 11.1).

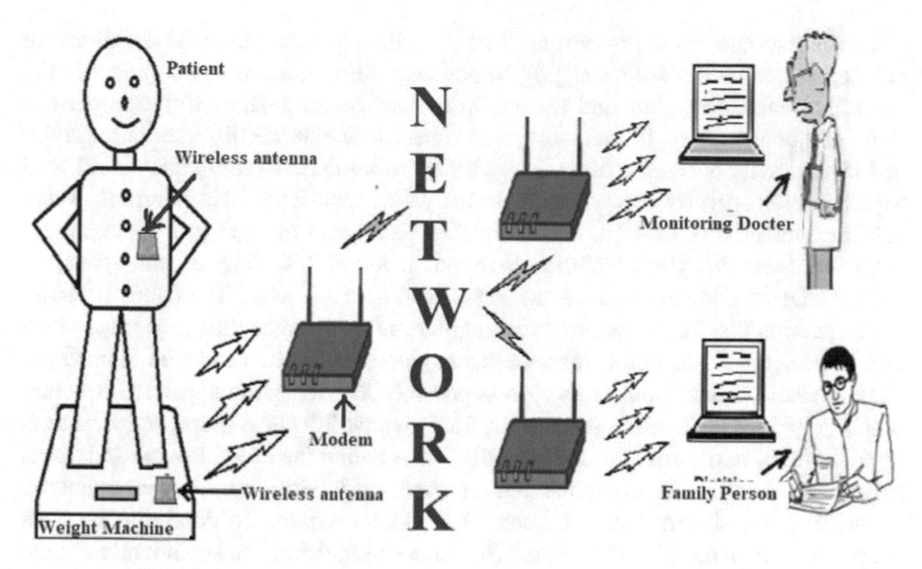

Fig. 11.1 Biomedical utilization of the recommend antenna

11.2 Related Work

Yang [1] In this study, a singular supported wideband implantable reception apparatus which works at the range of 2.4 GHz is planned for biomedical purposes. The impedance data exchange limit update is cultivated, by solidifying two modes. One is the significant strategy for centre microstrip antenna, and another available mode is accomplished by outer circle. Reproduced impedance information exchange limit and radiation plans are examined in spectre. The identical impedance information exchange limit underneath −10 dB degrees 2.24–2.59 GHz. The recommended reception apparatus, which has a zenith expansion of −20.8 dB, generally transmits in the off-body heading. The effect of $\in_r$ and install significance are furthermore discussed. The sensitivity ponders exhibit the robustness of our antenna. A single empowered implantable wideband radio wire is shown for 2.4 GHz biomedical purpose. The repeated S11 exchange speed underneath −10 dB ranges 2.24–2.59 GHz. The impedance data exchange threshold refinement is cultivated by integrating two modes. One notable strategy for centre MPA, and other is cultivated by the outer circle. The point of confinement of the recognized increment of minimum mode is gotten as −20.8 dB. Availability, the looking at estimation of max mode is −26.4 dB [4–8]. At last, the sensitivity examination of the cells shows the robustness of the radio wire.

Sinan [2] Chest fatal development impacts various women and has fatal consequences in case it doesn't fix properly. Early detection isprimitive variable to recognize and interfere with harmful development cell. Some of strategies for chest malady acknowledgment are mammography, X-ray and ultrasound. In any case, there are a couple of imperatives. For example; some place in the scope of 5 and 33% of all chest illnesses are missed by virtue of poor perilous/pleasant harmful development cell separation. Microwave imaging to perceive chest harmful development is the most favourable procedure and various work here. Materials have differing permittivity and conductivity. In that work, a 3D chest structure has different permittivity and conductivity is exhibited in HFSS 15.v by using FEM to comprehend EM field regards, microstrip fix reception apparatus working at 2.45 GHz is arranged and substrate material is FR4 ($\in_r = 4.4$ F/m). Opening on microstrip fix and changing ground seat, imaging resolution isextended. About this, electric field, appealing field scattering and stream thickness on antenna are assessed. The paper is about, the inspection that inset supported rectangular microstrip reception apparatus structure is investigated to outfit microwave imaging technique in order to dismember chest destructive development as soon as possible. The reception apparatus structure working at 2.45 GHz are imitated with fundamental 3D chest shape and structure. Specific radio wire structures are examined by varying the ground plane and opening on microstrip fix. Accomplishment of work, for the fourth radio wire structure, electrical field, alluring field and stream thickness regards on condition of chest shape with tumor are 137.37 V/m, 0.796 A/m ve 54.956 A/m^2 independently, for the properties in the situation of chest shap without the presence of tumor 170.38 V/m, 0.84534 A/m ve 68.152 A/m^2, independently. It will generally be stated that this work achieves better success rate when appeared differently in relation to works recorded as a hard copy. Depending upon the propagation outcomes and graphical observation, the fourth antenna structure gives the best revelation to the chest harm.

Sukhija [3] In this paper, a U-framed micro-strip fix radio wire with wound openings is introduce. It is planned for biomedical purposes to work at 2.45 GHz. In light of the diversion involvement in, two plans of the fix are given and without usage of meandered openings. The nearby examination between these two is likewise plot. It is seen that the antenna with wound openings shows extraordinary execution with adequate information exchange limit, low incidents and is fit for use in biomedical purposes. Available, the recommend reception apparatus have minimal size of $35 \times 29 \times 1.6$ mm^3, and the range of the ground plane is only 15% of the general radio wire gauge. The conscious and imitated results show incredible simultaneous with each other. The reception apparatus is made on a FR4 epoxy substrate, and propagation is done on FDTD placed Realm excel test framework. This paper reports an arrangement of U-shaped fix reception apparatus with meandered openings for ISM band application and elucidates the examination between exploratory outcomes of the recommend antenna in free space and muscle show. As a result of its wide information transmission incorporates, this radio wire can be used for remote prosperity watching structure. The purposeful and recreated results give similar response. Its diminutiveness, incredible repeat versus adversities response, radiation precedent will help this organized antenna with being a suitable contender for biomedical purposes. The depiction of the recommend antenna in checking machine will be of extraordinary interest. Future degree furthermore fuses more diminishing in size and SAR figuring of the recommend radio wire.

11.3 Antenna Design

Our goal is to design a wide-band accepting wire working at ISM band covering from 2.45 GHz for biomedical purpose. Schematic of the recommend radio wire is outlined in Fig. 11.2. The point by point parameters are recorded in Table 11.1. The recommend gathering device is a micro-strip fix accepting wire along a minuscule stick. The range of this short stick is 0.17 mm. A circle is printed exclude the MPA. The substrate are made of Rogers RT/duriod ($\epsilon_r = 2.2$, $\tan\delta = 0.0023$) with a thickness (h) of 1.6 mm.

For accommodation, the implantable focal point antenna is fixed as the inception of the organize framework as showed in Fig. 11.2. As would be clear from this, a 50 coax test is acquainted with sustain this gathering mechanical assembly direct.

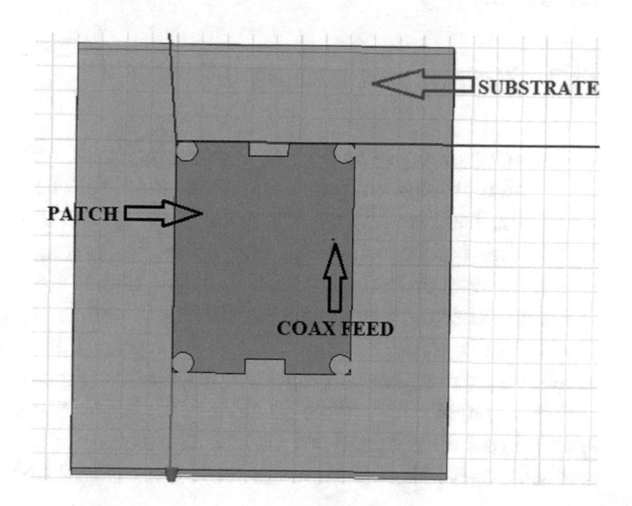

Fig. 11.2 Geometry of recommend antenna

Table 11.1 Parameter of antenna

Parameter	Dimensions (mm)
L_S	88
W_S	76
L_P	48
W_P	36
L_G	88
H_S	1.6
P_H	1.6
P_R	0.17
S_S	0.6

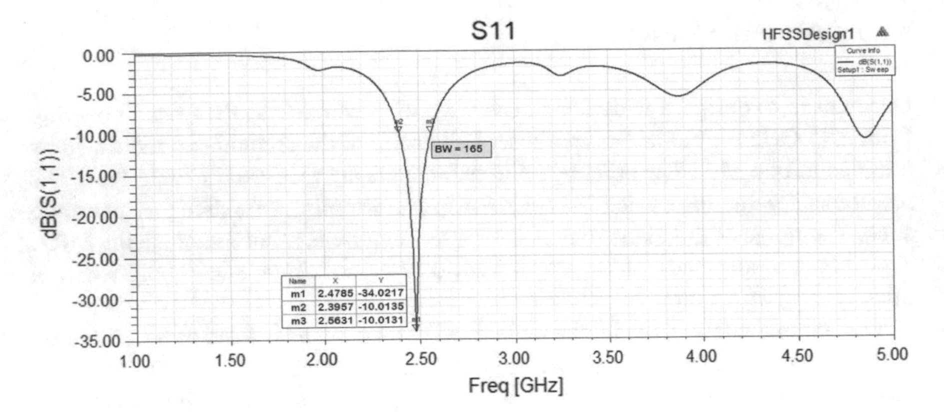

Fig. 11.3 S11 graph

11.4 Results and Discussion

The variety of antenna S11 as a component of recurrence mimicked by HFSS 15.v which utilizes limited components strategy. The reverberation frequencies are 2.45 GHz. Their individual dimension of the arrival misfortunes is −34.04 dB. Truth be told, this can be utilized for biomedical ISM band applications.

For the recommend antenna, 2 principle planes are picked to exhibit the radiation structure. These are suggested as the E&H planes. The radiation structures in the H-plane (x–z plane) and the E-plane (y–z plane). It may be seen that the radiation structures in x–z plane present two projections (Figs. 11.3, 11.4 and 11.5).

The expansion has some soundness in the reenactment repeat band and has an apex estimation of 6.8 dB at 2.4 GHz. The expansion is commonly extraordinary and can be improved by inserting our radio wire (Fig. 11.6).

11.5 Conclusion

This paper reports a plan of rectangular melded wideband antenna has been exhibited for 2.4 GHz ISM biomedical purpose. The re-enacted S11 data transmission underneath −10 dB extents from 2.45 GHz. Because of its wide data transmission includes, this antenna can be utilized for remote wellbeing observing framework. The limit of the acknowledged increase is gotten as −34.04 dB. Its smallness, great recurrence versus misfortunes reaction, radiation example will assist this structured antenna with being a successful contender for biomedical purposes. The portrayal of the recommend antenna in gauging machine will be of extraordinary intrigue. Future extension availability incorporates more decrease in size and SAR estimation of the recommend antenna.

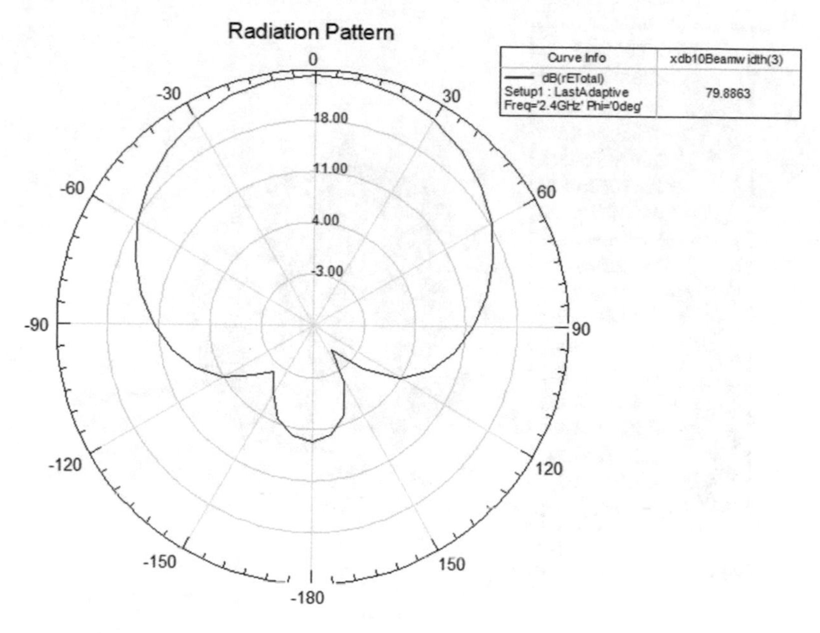

Fig. 11.4 Radition pattern

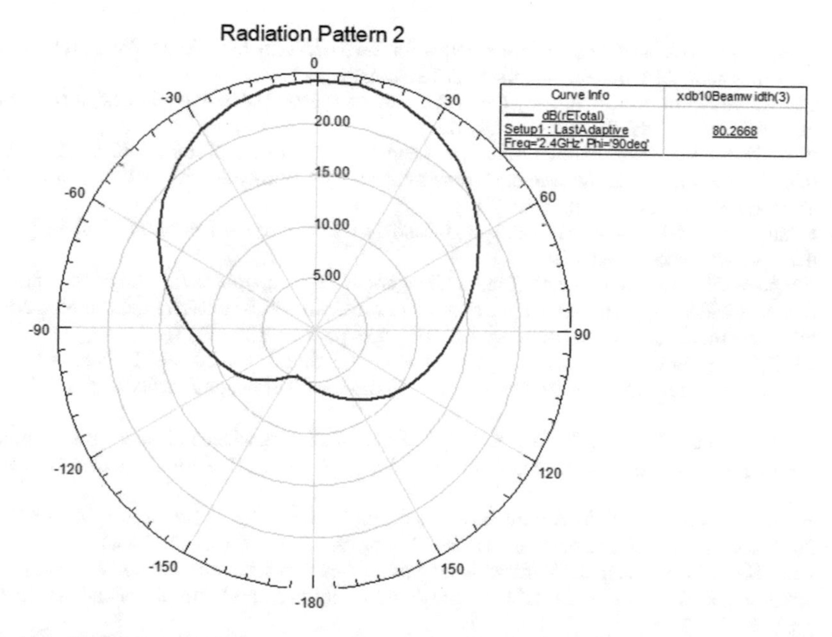

Fig. 11.5 Radition pattern

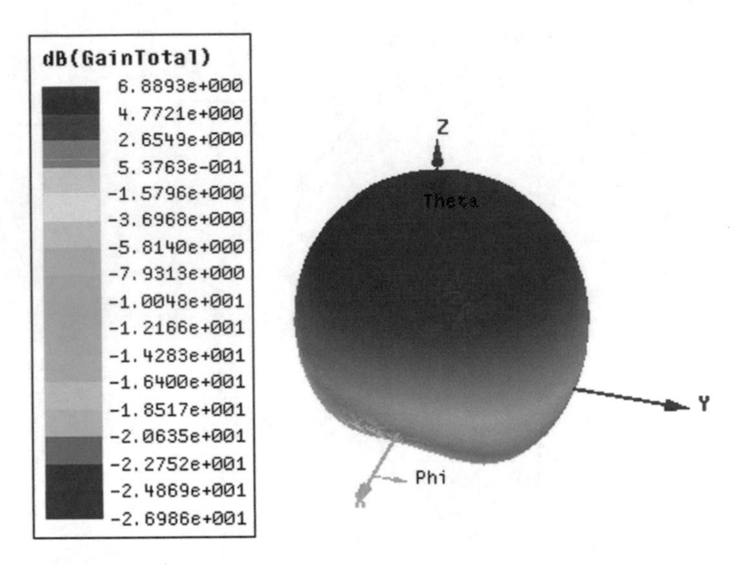

Fig. 11.6 Gain

References

1. Yang, Z.: A wideband implantable antenna for 2.4 GHz ISM band biomedical purpose. In: 2018 International Workshop on Antenna Technol, pp. 1–3
2. Çal, R., Gültekin, S.S., Uzer, D., Dündar, Ö.: A microstrip patch antenna design for breast cancer detection. **195**, 2905–2911 (2015)
3. Aziz, N.A., Mohamad, N.R., Abu, M., Othman, A.: Design of ultra-wideband (UWB) implantable antenna for biomedical telemetry design of ultra-wideband (UWB) implantable antenna for biomedical telemetry (2016)
4. Sukhija, S., Sarin, R.K.: A u-shaped meandered slot antenna for biomedical purposes. **62**(November), 65–77 (2017)
5. Abufanas, H., Hadi, R.J., Sandhagen, C., Bangert, A.: New approach for design and verification of a wideband archimedean spiral antenna for radio-metric measurement in biomedical purposes. In: 2015 German Microwave Conference, pp. 127–130 (2015)
6. Kwak, S., Chang, K., Yoon, Y.J.: Ultra-wide band Spiral shaped small antenna for the biomedical telemetry. In: 2005 Asia-Pacific Microwave Conference Proceedings, vol. 1, p. 4 (2005)
7. Chen, X., Liang, J., Wang, S., Wang, Z., Parini, C.: Small ultra wideband antennas for medical imaging (invited paper). In: Chen, X., Liang, J., Wang, S., Wang, Z., Parini, C. (eds.), pp. 28–31 (2008)
8. Ruvio, G., Ammann, M.J.: A miniaturized antenna for UWB-based breast imaging. In: 2009 3rd European Conference on Antennas and Propagation, pp. 1864–1867 (1864)
9. Lam, H.H., Bornemann, J.: Ultra-wideband printed-circuit array antenna for medical monitoring applications. In: 2009 IEEE International Conference on Ultra-Wideband, vol. 2009, pp. 506–510 (2009)
10. Clemente, F.S., Helbig, M., Sachs, J., Schwarz, U., Stephan, R., Hein, M.A.: Permittivity-matched compact ceramic ultra-wideband horn antennas for biomedical diagnostics. In: Proceedings of the 5th European Conference on Antennas and Propagation, pp. 2386–2390 (2011)
11. Sabban, A., Member, S.: New wideband printed antennas for medical applications. IEEE Trans. Antennas Propag. **61**(1), 84–91 (2013)

12. Sabban, A., Member, I.S.: New wideband tunable printed antennas for medical applications. In: Proceedings of the 2012 IEEE International Symposium on Antennas and Propagation, p. 1–2 (2012)
13. Scott, F., Stephan, R., Hein, M.A.: Ultra-wideband miniaturised permittivity-matched antennas for biomedical diagnostic. In: 2013 7th European Conference on Antennas and Propagation, pp. 2896–2899 (2013)
14. Khaleghi, A., Balasingham, I., Vosoogh, A.: A compact ultra-wideband spiral helix antenna for in-body communications.In: The 8th European Conference on Antennas and Propagation (EuCAP 2014), pp. 3093–3096 (2014)
15. Ahmed, F., Hasan, N., Chowdhury, H.M.: A compact low-profile ultra wideband antenna for biomedical purposes. In: 2017 International Conference on Electrical, Computer and Communication Engineering, pp. 87–90 (2017)

Chapter 12
Implementation of MRI Images Reconstruction Using Generative Adversarial Network

Mahesh Pawar, Sandeep Kakde and Prashant Mani Yadav

Abstract This paper focuses on a Magnetic Resonance Imaging (MRI) reconstruction that gives brisk realization. This diminishes the scanning cost and image reconstructed in very fewer time. In this method, Generative Adversarial Network (GAN) designed a generator which gives the better enhancement like texture smoothness, and high resolution. In addition, it also finds the frequency province information to embed resemblance in both the images using parameter Structural Similarity Index (SSIM). Also performed radon transform to find the structural similarity of images with enhancement, accuracy and test whether the images are real or fake. Compared to other methods, the proposed GAN method provides superior reconstruction.

Keywords Image processing · Generator · Discriminator · PSNR

12.1 Introduction

Magnetic Resonance Imaging (MRI) is mostly used for scanned imaging application. MRI scans the body tissue and gives excellent contrast, which also includes the structural and functional information of whole body. The main drawback of MRI is slightly slow speed because data samples cannot directly collected in an image, but rather than in specific area. This data has special time period information, it collects data serially. For high quality of image up to 512 lines of data needs. During MRI patients movement and other physiological motion gives slow speed. MRI data samples are obtain sequentially in k-space. K-space determined by Nyquist-Shannon sampling criteria.

M. Pawar (✉) · S. Kakde
Electronics Engineering Department, Y C College of Engineering,
Nagpur University, Nagpur, India
e-mail: mahesh2323@gmail.com

S. Kakde
e-mail: sandip.kakde@gmail.com

P. M. Yadav
Electronics and Communication Engineering Department, SRM University, Ghaziabad, India
e-mail: prashanm@srmist.edu.in

V. E. Balas et al. (eds.), *Internet of Things and Big Data Applications*, Intelligent Systems Reference Library 180, https://doi.org/10.1007/978-3-030-39119-5_12

12.2 Previous Work

Mardani et al. proposed an novel compressed sensing framework for faster and more valuable image reconstruction [1] from under sampled data. Reconstruction made on historical data for higher resolution sharp and effective contrast. Uses a LSGAN for training generator which contains the under sampled images for reducing aliasing artifacts. Oktey, Bai et al. describes cardiac image excellent resolution by means of Convolutional Neural Network [2]. Multiple data takes from different plane for improved performance cardiac short and long axis magnetic resonance images gives output as CNN approach well performed state-of-art super resolution (SR) method. Also perform segmentation and motion tracking benefits. Pham et al. focuses on SRGAN, for image great resolution using a GAN methos [3]. Develop a framework for photo realistic neural image for 4× up scaling factor. To achieve this uses a perceptual loss and content loss presented in adversarial loss. This is the best solution to train the image using discriminator network and differentiate between super resolved images and original photo-realistic images. Ledig et al. proposed a deep Three Dimensional (3D) convolutional neural network (CNN) for high resolution of brain images [4]. Extend super resolution images to multimodal super resolution using intermodality.

Jia et al. proposed a algorithm multi-frame super-resolution (SR) reconstruction based on sparse representation of medical MR images corresponding data between high resolution (HR) slices and low resolution (LR) section scans also find the self similarity between then to up-sample the input scan [5]. Du et al. worked on reconstruction of visual stimulus of missing view in different view variable structure shares a common latent representation using generative adversarial network extracts the nonlinear features from visual data and correlation between voxel activities of FMRI recordings [6–10]. These activities are very tough for removing noise and improving prediction. 3 FMRI recording are more precisely reconstruct [11–18].

12.3 Proposed Methodology

This paper offers a framework which depends on Generative Adversarial network (GAN) for reconstruction and resolution as shown in Fig. 12.1. GAN provide a very effective framework for generating MRI images with elevated perceptual quality. This method promotes reconstruction to progress closer to the investigate space with probability.

A. Generative Enhancement

In the first step of generative enhancement, collections of MRI images are stored into the database. For enhancement of scanned MRI image, step wise generation are applied on image to get perceptual quality image and performs smoothing to

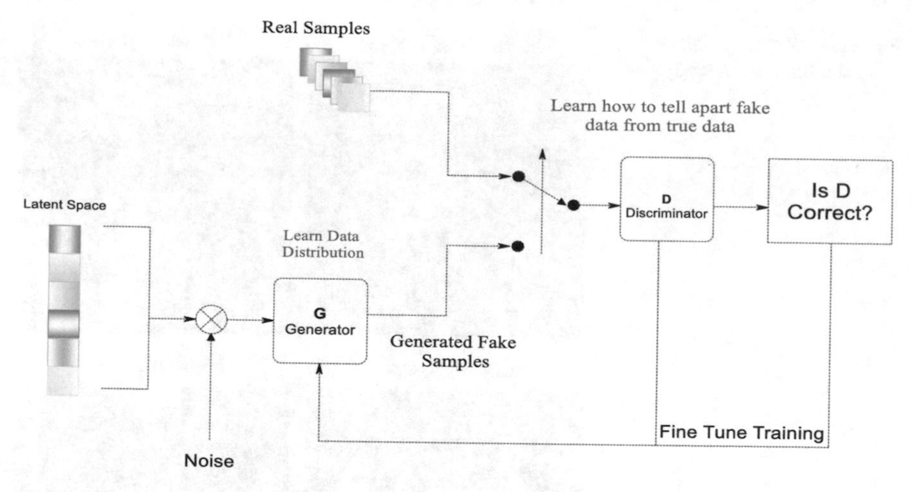

Fig. 12.1 GAN for fast CS-MRI

reduce noise, noise is anything in the image unwanted or unnecessary information. Figure 12.2 shows images using Gaussian Filter. Using Laplacian Gaussian filter, smoothing is perform to reduce noise in image. Due to presence of noise in image there is some improper data, with the help of supervised learning boosts the minor areas to get expected data, finally for intensity level clipping performed. After applying all parameters on image enhanced MRI image is reflected.

B. Image Segmentation

Key parameters for analysis of an images are color moments, texture or edge. Segmentation change the representation of an image that is more meaningful, classifies the pixels into object or group and easy to analyze. In this paper, measurement is taken in k-space i.e., the value of k-space is 3 (K = 3), k-space has spatial frequency information and it collects data line by line. K-space differentiated into three regions namely White Region, Gray Region and Fluid Region. Double Thresholding algorithm is used for removing unnecessary detail and brings out hidden details. Enhanced

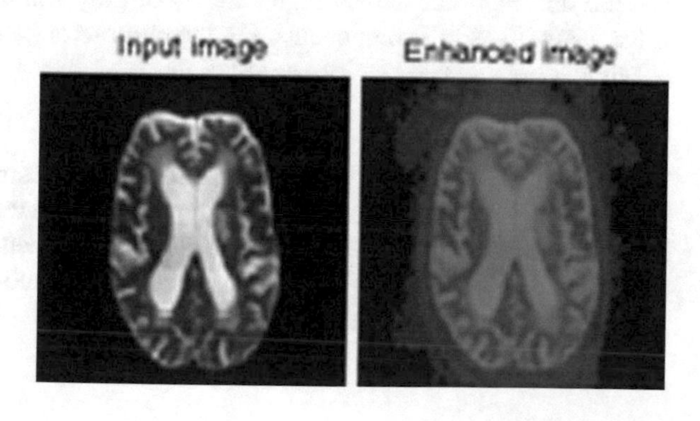

Fig. 12.2 Enhanced image using gaussian filter

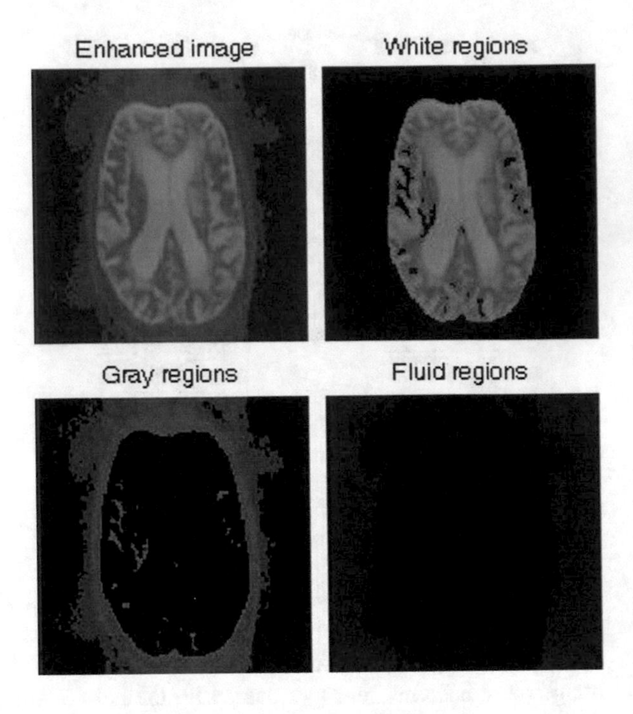

Fig. 12.3 Segmentation using double thresholding

image is group of shape, texture and other information. Separated the white, gray and fluid region for analysis of further process to get a better image. Figure 12.3 shows segmentation using Double Thresholding method.

C. Discriminator

Edge Detection.

Feature selection is one of the most important segments in this research. The main goal of feature extraction is to convert segmented objects into representation that better describes their main features and attributes. In this paper, feature extraction used for locating areas with strong intensity contrast. It recognizes objects boundaries also extract corner, lines and curves. Figure 12.4 shows edge features of white components and Fig. 12.5 shows edge features of gray components. Figure 12.6 shows Edge features of CSF components. For the purpose of better resolution, edge feature of white, gray and CSF regions are finding.

D. Projection and Radon Transform

The radon transform is an integral transform for reconstruction of medical scanned images. Inverse of radon transform helps to reconstruct the medical images. It reconstruct the object from projection data, it takes one dimensional fourier transform of projection data for deriving radon transform. The projection data calculate the object using two dimensional inverse-fourier transform.

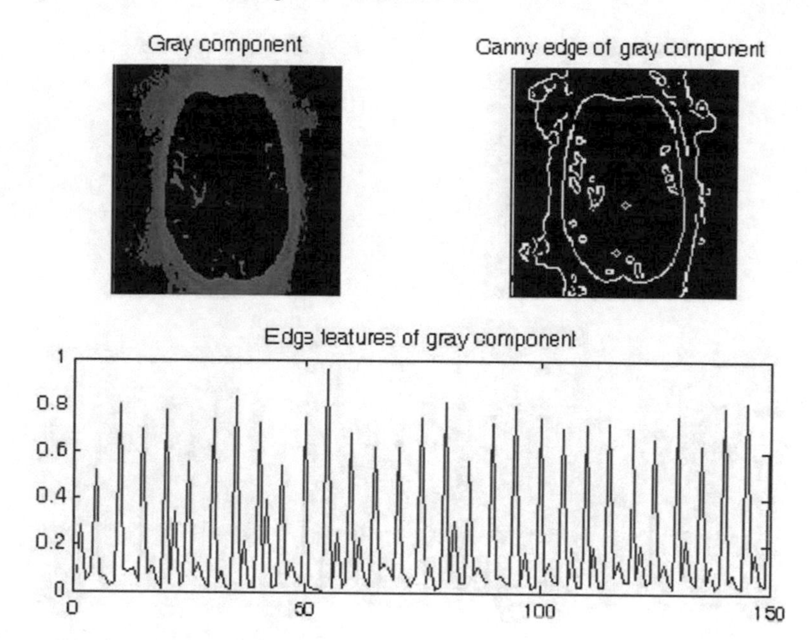

Fig. 12.4 Edge features of white components

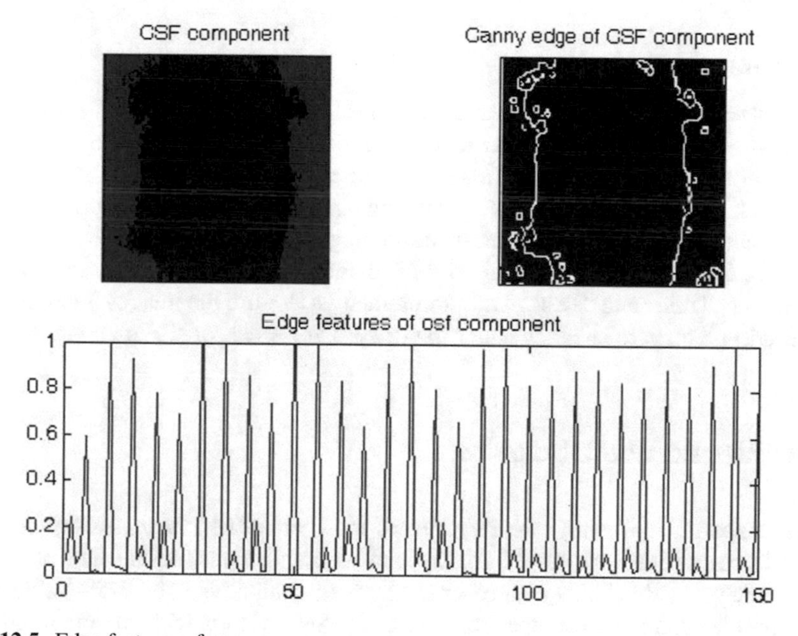

Fig. 12.5 Edge features of gray components

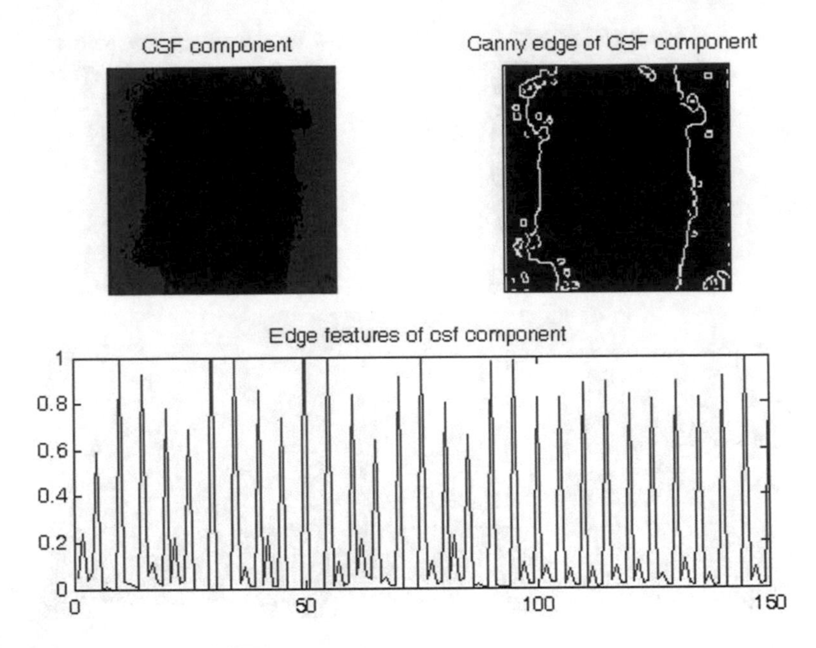

Fig. 12.6 Edge features of CSF components

E. Performance Evaluation

K-NN classifier is most widely used classifier in image processing. If numbers of samples are large, classification is much better. In this paper, K choose as a 2 i.e., taking two input images. All related images are stored into database; according to stored database k classifies the new object based on similarity index. Structural Similarity Index (SSIM) of image obtained after subtracting 'one' from dissimilarity in images. Figure 12.7 shows Structural similarity of image 1 and 2. The two parameters are: implement "Train" and "Test". Train implement is to store the images into database and test implement to check whether the images are 'Real' or 'Fake'.

12.4 Results and Discussion

In this experimental study, we reconstruct the brain MRI images using Generative Adversarial Network (GAN) for fast and accurate reconstruction. The proposed method suggests that GAN beat other methods in qualitative validation. Table 12.1 tabulates the quantitative comparison result of GAN variation with other method and overall table shows the improved MSE and PSNR.

We address MSE, PSNR and structural similarity; also evaluate qualitative visualization of reconstructed MRI images. Table 12.2 shows the performance evaluation of images with 90% accuracy and structural similarity (Fig. 12.8).

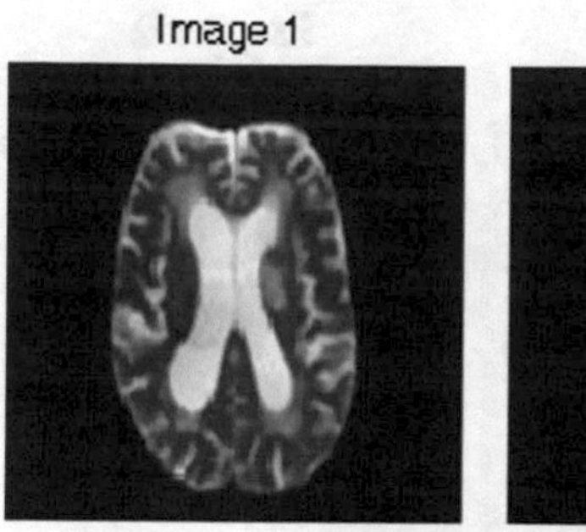

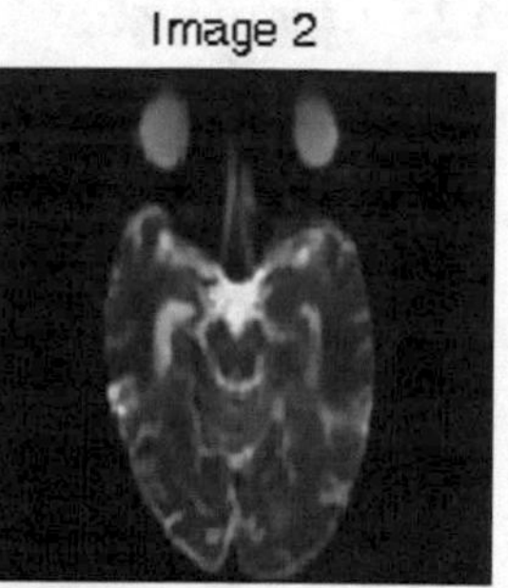

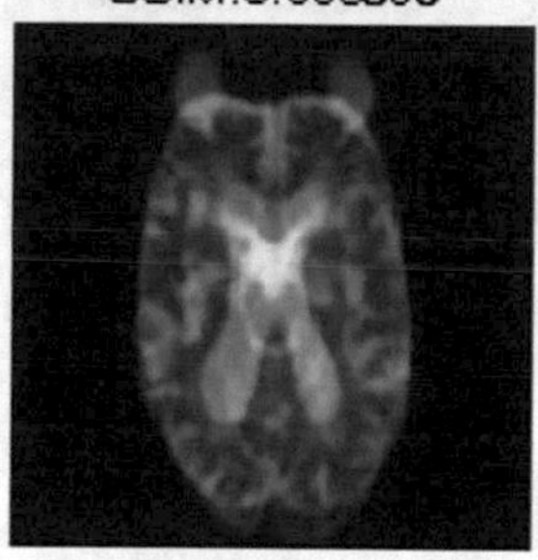

Fig. 12.7 Structural similarity of image 1 and 2

Table 12.1 Comparison with other techniques

S. No.	Methods	MSE	PSNR (dB)
1	PG	0.05	44.39
2	PPG	0.05	45.61
3	PPGR	0.04	47.30
4	PFPGR	0.04	47.83
5	GAN–White region Gray region CSF region	0.0023 0.0030 0.0031	50.5203 49.2901 49.1679

Table 12.2 Performance evaluation of images

Images	Actual image	Obtained image	SSIM
1	Real	Real	0.9983
2	Real	Real	0.9976
3	Fake	Fake	0.9975
4	Real	Real	0.9983
5	Fake	Fake	0.9982

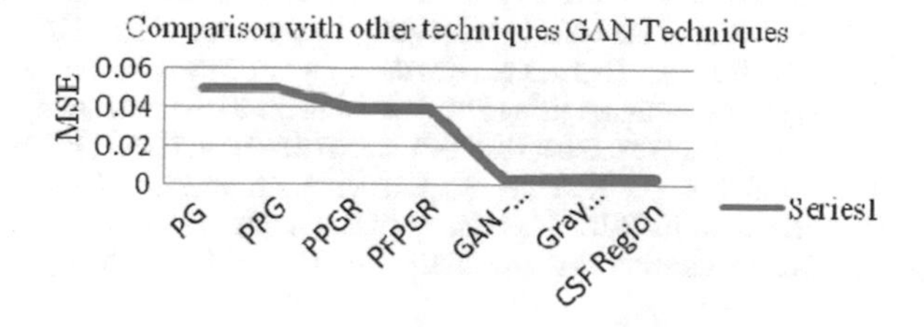

Fig. 12.8 Comparison with other GAN techniques

12.5 Conclusion

In this paper, the proposed method is easy to find the structural similarity of images than other methods. MSE and PSNR are the design parameters for measurement and test this method. We compared following DAGAN variation. The higher PSNR values shows more image enhancement. We achieved higher PSNR (i.e., PSNR > 48). With this technique, 90% accuracy is achieved in images.

References

1. Mardani, M., Gong, E., Cheng, J.Y., Vasanawala, S., Zaharchuk, G., Alley, M., Thakur, N., Han, S., Dally, W., Pauly, J.M., Xing, L.: Deep generative adversarial networks for compressed sensing automates MRI. arXiv Prepr (2017)
2. Oktay, O., Bai, W., Lee, M., Guerrero, R., Kamnitsas, K., Caballero, J., de Marvao, A., Cook, S., O'Regan, D., Rueckert, D.: Multi-input cardiac image super-resolution using convolutional neural networks. In: Medical Image Computing and Computer-Assisted Intervention (2016)
3. Pham, C.H., Ducournau, A., Fablet, R., Rousseau, F.: Brain MRI super-resolution using deep 3D convolutional networks. In: 2017 IEEE 14th International Symposium on Biomedical Imaging (ISBI 2017) (2017)
4. Ledig, C., Theis, L., Huszar, F., Caballero, J., et al.: Photo-realistic single image super-resolution using a generative adversarial network. Preprint at arXiv:1609.04802 [cs.CV] (2016)
5. Jia, Y., Gholipour, A., He, Z., Warfield, S.K.: A new sparse representation framework for reconstruction of an isotropic high spatial resolution mr volume from orthogonal anisotropic resolution scans. IEEE Trans. Med. Imaging **36**(5), 1182–1193 (2017)
6. Du, C., Du, C., He, H.: Sharing deep generative representation for perceived image reconstruction from human brain activity. In: International Joint Conference on Neural Networks (IJCNN) 2017, pp. 1049–1056 (2017)
7. Korus, P.: Digital image integrity—a survey of protection and verification techniques. Digit. Signal Process. **71**, 1–26 (2017)
8. Wen, D., Han, H., Jain, A.K.: Face spoof detection with image distortion analysis. IEEE Trans. Inf. Forensics Secur. **10**(4), 746–761 (2015)
9. Hollingsworth, K.G.: Reducing acquisition time in clinical MRI by data undersampling and compressed sensing reconstruction. Phys. Med. Biol. **60**(21), 297–322 (2015)
10. Nguyen, H.M., Glover, G.H.: A modified generalized series approach: application to sparsely sampled FMRI. IEEE Trans. Biomed. Eng. **60**(10), 2867–2877 (2013). https://doi.org/10.1109/TBME.2013.2265699
11. Bhanuse, S.S., Kamble, S.D., Kakde, S.M.: Text mining using metadata for generation of side information. Procedia Comput. Sci. **78**, 807–814 (2016)
12. Humerah, N., Rathkanthiwar, S., Kakde, S.: Implementation of hybrid algorithm for image compression and decompression. Int. J. Eng. Res. **5**(5), 398–403 (2016)
13. Hatwar, R.B., Kamble, S.D., Thakur, N.V., Kakde, S.: A review on moving object detection and tracking methods in video. Int. J. Pure Appl. Math. **118**(16), 511–526 (2018)
14. Awaghate, A., Thakare, R., Kakde, S.: A brief review on: implementation of digital watermarking for color image using DWT method. In: 2019 International Conference on Communication and Signal Processing (ICCSP), pp. 0161–0164. IEEE (2019)
15. Nirmalkar, N., Kamble, S., Kakde, S.: A review of image forgery techniques and their detection. In: 2015 International Conference on Innovations in Information, Embedded and Communication Systems (ICIIECS), pp. 1–5. IEEE (2015)

16. Shende, P., Pawar, M., Kakde, S.: A brief review on: MRI images reconstruction using GAN. In: 2019 International Conference on Communication and Signal Processing (ICCSP), pp. 0139–0142. IEEE (2019)
17. Channe, R., Ambatkar, S., Kakde, S., Kamble, S.: A brief review on: Implementation of lossless color image compression. In: 2019 International Conference on Communication and Signal Processing (ICCSP), pp. 0131–0134. IEEE (2019)
18. Karras, T., Aila, T., Laine, S., Lehtinen, J.: Progressive growing of GANs for improved quality, stability, and variation. In: The International Conference on Learning Representations (ICLR) (2018)

Chapter 13
An Emerging Approach to Intelligent Techniques—Soft Computing and Its Application

Pankaj Kumar and S. K. Singh

Abstract Over the past decade, artificial intelligence (AI) has appeared in high-tech labs, and people use it daily without feeling it. For example, current calculations of power technologies and flexible computing in other areas related to digital products, many applications that increase the contribution of artificial intelligence, are flexible computing technologies based on these problems, the medicine, biology, industry, manufacturing described as security, education, virtual environment. We was particularly encouraged. Support for the provision of new ideas and/or theoretical studies on soft computing technology, interdisciplinary research, real-world experimental application problems and flexible computer system design descriptions, diverse in development their new analytical structure Introduce the document Encourage a soft computing technology that is particularly practical. It provides an overview of the disciplines of Artificial Intelligence with respect to Soft Computing, and presents the benefits of soft computing over traditional hard computer technologies and their disadvantages.

Keywords AI · Neural network · Fuzzy logic · Genetic algorithms · Neuron

13.1 Introduction

At the beginning of the Stone Age, when people started fleeing into the caves, they attempted to immortalize themselves by painting their image on the rock. With the progressive progress of civilization, they were interested in seeing themselves in different ways. That is why they took millions of years to construct an analytical engine, which can earn millions of mechanical arithmetic. The Babbage analytical

P. Kumar (✉)
Department of Computer Science & Engineering, NIET, Greater Noida, Uttar Pradesh 201306, India
e-mail: unpankaj@gmail.com

S. K. Singh
Department of Computer Science & Engineering, GCET, Greater Noida, Uttar Pradesh 201306, India
e-mail: singhsks123@gmail.com

V. E. Balas et al. (eds.), *Internet of Things and Big Data Applications*, Intelligent Systems Reference Library 180, https://doi.org/10.1007/978-3-030-39119-5_13

engine is the first big hit in the modern computer age. The 5th generation computer is equipped with a high-speed VLSI engine, can detect objects with cameras and place it at the desired location. During 90s, the Japanese government start producing the fifth generation computer. It can process natural language, prediction, fault tolerance, massively parallel processing, progressive degradation, inductive learning, adopted environmental noise data or data discourse of the 5th generation computer, the image of the cognitive object and the mathematical theorem are under this soft computing. the real world, parallel and distributed information included in the field of artificial intelligence It can be integrated with the information processing and the basic units, at higher levels of connectivity, but what is artificial intelligence?

Artificial intelligence, like intelligent human thinking, is a way to intelligently think of a computer, a computer-controlled robot or software. Artificial intelligence is achieved by studying how the human brain thinks, learns, solves and works when trying to solve problems. The researchers defined various definitions (Table 13.1).

These definitions provide a specific framework for AI, but can be short-lived as the conceptual framework evolves rapidly. The AI can be divided into nine special branches.

- *Problem Solving and Planning*: This objective is related to hierarchical system depreciation, planned review mechanisms, and centralized search for essential purposes.
- *Expert system*: Associated with the processing of knowledge and the adoption of complex decisions.
- *Natural language processing*: Automatic text generation is associated with text processing, machine translation, speech synthesis and analysis, grammatical analysis and style.

Table 13.1 Definition of AI given by researchers and authors

Year	Authors	Definition
1956	John McCarthy	The science and engineering of making intelligent machines, especially intelligent computer programs
1987	M. Mitchell Waldrop	AI is the art of making computers do smart things
1988	Raymond McLeod	AI is the activity of providing such machines as computers with the ability to display behavior that would be regarded as intelligent if it were observed in humans
1989	William A. Taylor	AI is a programming style, where programs operate on data according to rules in order to accomplish goals"
1991	Rich & Knight	The study of how to make computers do things at which, at the movement, people are better
1993	Luger & Stubblefield	AI is a branch of computer science that is concerned with the automation of intelligent behavior
2003	Stuart Russell and Peter Norvig	AI is the study of agents that exist in an environment, perceive, and act

- *Robots*: Control robots, manipulate objects, and use information from sensors to guide actions.
- *Computer vision*: Intellectual visualization, visual analysis, understanding and processing of images and speed traversal.
- *Learning*: This machine is related to the research and development of different forms of learning.
- *Neural Network*: combination of simulated recognition of human brain models and derivable numerical and logical calculations.
- *Fuzzy Logic*: This theory of fuzzy sets attempts to capture human representation and logic in an uncertain state through knowledge of the real world. Uncertainty may arise due to generality, ambiguity, ambiguity, opportunity, or incomplete knowledge.
- *Genetic Algorithms*: These are adaptive algorithms with potential learning opportunities. They are used for research, machine learning and optimization.

We have found that complex real problems require an intelligent system combining knowledge, techniques and methods from different sources. These intelligent systems are considered experts in specific areas, how to better optimize and modify the environment, and determine behaviors. In the real world of computer problems, it is often advantageous to use a variety of computer technologies, particularly to build intelligent systems using complementary hybrids acquired sequentially rather than sequentially. The best way to design such a smart system is to group together different types of soft computing branches. Creating an innovative approach to soft computing has become an important news.

The soft computing is useful for traditional computing comparisons, estimation models and complex computer problems. Unlike hard computing, soft computing allows opacity, uncertainty, partial authenticity and prediction. This method can solve problems that may or may not be necessary to solve common hardware problems.

Soft computing combines technologies such as artificial neural networks, fuzzy logic, and genetic algorithms. Today, soft computing technology has been successfully applied in many household, commercial, and industrial applications with the advent of high-performance, low-cost digital processors and low-memory chips. The soft computing model is the human mind.

The rest of this paper is structured as follows. Section 13.2 describes the computing and their types. Section 13.3 is devoted to hard computing, and Sect. 13.4 is about soft computing. Section 13.5 illustrates the difference between hard and soft computing. Section 13.6 describes soft computing technology/components. Section 13.7 describes the use of soft computing. Section 13.8 discusses the future of software computing, and Sect. 13.9 summarizes this paper.

13.2 Computing

Computer science concerns the use of computers and computer systems, as well as their creation and programming. The main aspects of his theory, systems and applications are topics of technology, design, engineering, mathematics, physics and social sciences.

The field of engineering sciences investigating the process and structure with the help of computers is known as computing [1, 2].

There are two types of computing:

(a) Hard computing
(b) Soft computing.

13.3 Hard Computing

Hard computing [2, 3] based on the principles of accuracy, uncertainty and stiffness are based on mathematical or analytical models, binary logic, transparent systems, numerical analysis and transparent software, and require high precision. It has a set of predefined rules. This requires a quick limit, so the calculation takes a lot of time. This is called traditional computers. Example: Arithmetic sum, Newton's law, etc. This is related to the precise model of the exact solution, as shown in Fig. 13.1.

13.3.1 Advantages

- It is linked to precise calculations.
- The rules of hard computation are strict and binding.
- The goal solution is the process of every problem. All input data, products and processes are clearly defined.
- Most laws of physics, mathematics and exact sciences are written using rigorous calculation principles.

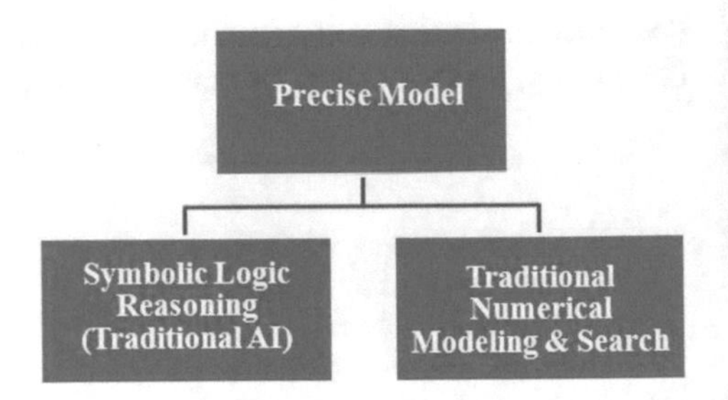

Fig. 13.1 Precise model

13.3.2 *Disadvantages*

- It is difficult to control real problems using computational methods that make it difficult to change information or manage behavioral barriers.

13.4 Soft Computing

The inventor of fuzzy logic, Lotfi A. Zadeh, proposed the term "soft computing". They defined as "a compilation of flexible computing functions designed to take advantage of tolerance, mistrust and uncertainty to achieve management, reliability and low-cost solutions."

Soft computing [1–3] is an industry created using intelligent and complex machines. This new method of computing allows for calculations that are comparable to the ability of human thinking to study logic, uncertainty, truth and partial training, as well as the ability to cope with the costs of resolution and low resolution in the environment. This refers to the use of estimation models and offers an impossible, but useful solution to complex computational problems. The model is illustrated in Fig. 13.2.

Soft processing is based on a variety of biological methods, such as neural networks, fuzzy logic and genetic algorithms. Now that the problem has not been resolved using mathematical modeling (i.e. using an algorithm), the only solution to find is that it can be easily implemented in the case of complex problems, changing the nature of the problem. Real-time scene and parallel computing there are many applications with improved applications such as medical diagnostics, computer vision, handwriting, pattern recognition, artificial intelligence, weather conditions, network optimization, VLSI design and even better applications.

It has three main components shown in Table 13.2 with its strengths.

Unlike analytical methods, soft computational methods mimic consciousness and cognition in many important cases:

- They can learn from experience.
- They can be popularized in areas where there is no direct experience.

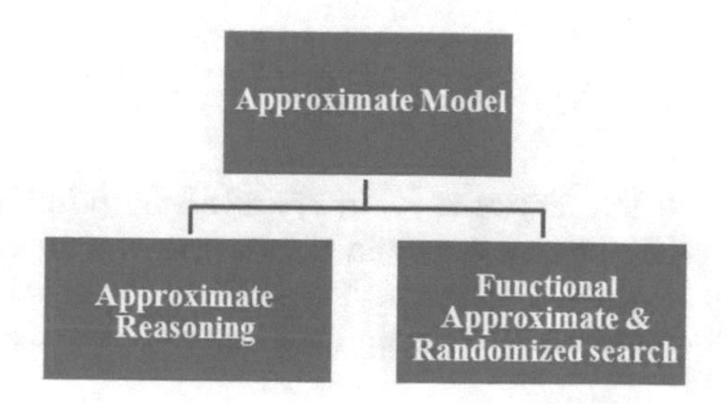

Fig. 13.2 Approximate model

Table 13.2 Components of soft computing

S. No.	Methodology	Strengths
1.	Neural network	Learning and adaption
2.	Fuzzy logic	Represent knowledge by fuzzy it-then rules
3.	Genetic algorithms	Random optimization for best results

- Simulation of parallel computer architecture of biological processes.
- They speed up I/O mapping with sequential analysis views.

13.4.1 Advantage

- Has a rich knowledge of representation (both at signal and pattern level).
- The process of acquiring knowledge is flexible (including machine learning and specialist training).
- It has a flexible knowledge processing.

13.5 Comparison Between Hard Computing and Soft Computing

Table 13.3 describes the comparison between Hard Computing and Soft Computing as follows.

13.6 Techniques/Components of Soft Computing

There are three main components of soft computing:

13.6.1 Neural Network

Neural networks [4, 5] are relatively primitive electronic models based on nerve structures of the brain. Learn from the experience of the brain. It is natural that some small energy saving solutions can solve some of the problems besides current computers. This brain modeling also ensures that there are few technical methods

Table 13.3 Hard computing versus soft computing

Hard computing	Soft computing
Hard computations requiring an analytical model to accurately describe traditional computations often require a lot of computational time	Soft computations differ from traditional (hard) computations in that they allow for irrationality, uncertainty, partial truth and approximation, as opposed to heavy computations. In fact, an example of easy computation is the human mind
It is based on binary logic, crisp systems, numerical analysis and crisp software	It is based on fuzzy logic, neural networks and probabilistic control
Although the calculations are complex, imprecision and uncertainties are undesirable qualities	Allowance for imprecision and uncertainty in soft computing can provide management capabilities, low cost, intelligent computing quotas, and savings in communications
We must write a program	We can develop own programs
These are two valuable arguments	We can use multi valuable or fuzzy logic
It is deterministic	It is stochasticity
This requires accurate input	It can process fuzzy and noisy data
It gives the exact answer	We can give an approximate answer

to develop mechanical solutions. This new calculation method leads to aesthetic degradation at system overloads comparable with similar conventional products.

Neural networks are trying to mimic the biological nervous system associated with creating therapeutic and informational strategies. These are parallel and decentralized information processing systems associated with loads and controlled by a biological education system, such as the human brain. In the neural network architecture, as a rule, there is a network of nonlinear information organized by levels and executed in parallel. This hierarchical network system is called the neural network topology. Nonlinear processing elements of this information in the network are defined as neurons, and the connections between these neurons in the network are called synchronization or weighting. Learning algorithms should be used to create neural networks so that they can process information in a meaningful and useful way. The network is trained to estimate the weights of the connections between the systems using the first suitable learning algorithm. After creating the network, we can sort the unknown test signals. The most common neural network multilayer network for classification functions is active [2, 4].

Most neural networks use well-formed learning algorithms. This means that the required results must be provided for each record used in the training. In other words, the inputs and outputs are known. During exercise monitoring, the network processes the input data and processes it using the results instead of the actual output. The error is then repeated on the network, and the load on the control network is compatible with the error returned. This process is repeated until the error is reduced, which means that only one data set is processed several times, because the weights between the network levels are adjusted during network training. This supervised learning algorithm, often called the core algorithm, is useful for studying interdisciplinary conceptual neural networks (MLP).

Neural networks are used in various applications, such as model classification, language processing, modeling, control, optimization and prediction of integrated systems. Neural networks have been used for analytical work in many applications in bioinformatics, such as predicting the secondary structure of a protein, gene expression profiles, and gene expression patterns by DNA sequences.

13.6.2 Fuzzy Logic

Fuzzy logic is considered similar to human thinking and the recognition of natural language. If this is ambiguous, things are not very clear or unclear. In real life, we may encounter situations in which it is impossible to determine whether a statement is true or false. Meanwhile, Fuzzy Logic offers more flexibility in arguments. We can also take into account the uncertainty of any situation [1, 6, 7].

After reviewing all available data, the fuzzy logic algorithms help to solve the problem. Members feel what they think about employees. Determines how each point in the input space is used to map the membership space. The subscription values in a fuzzy set are in the range [0:1] to make the best decision for a specific input. The Fuzzy logic method mimics the way people make decisions, taking into account all the possibilities between numbers 0 and 1 [8].

Features of Fuzzy Logic

- Machine learning technology is flexible and easy to implement.
- Human emotions can help us copy the logic.
- Very suitable infinite or approximate logical method.
- Fuzzy logic allows creating nonlinear functions of arbitrary complexity.

Advantages of Fuzzy Logic

- Fuzzy logic is widely used for commercial and practical purposes.
- Help to control machines and consumer products.
- It does not give exact reasoning, but the only acceptable reasoning.
- This can help us manage the uncertainty of our project.
- There is no need for accurate capture, because most are very powerful.
- Can be easily changed to improve or change system performance.
- It provides the most effective solution to complex problems.

Disadvantages of Fuzzy Logic

- Fuzzy logic is not always accurate, so the results are perceived based on assumption, so it may not be widely accepted.

- Fuzzy systems don't have the capability of machine learning as-well-as neural network type pattern recognition.
- Validation and Verification of a fuzzy knowledge-based system needs extensive testing with hardware.
- Setting exact, fuzzy rules and, membership functions is a difficult task.
- Some fuzzy time logic is confused with probability theory and the terms.

13.6.3 Genetic Algorithms

The genetic algorithm (GA) [9, 10] is an adaptive algorithm of discovery and optimization similar to the principles of natural selection and genetics. This is very different from the traditional search and optimization techniques used in various issues related to production and optimization. Thanks to its transparency, ease of use, minimum needs and global potential, GAs is proud that it can be used in all problem areas. GAs was developed by John Holland at the University of Michigan in 1965 and is inspired by Darwin's theory of development.

GA starts with a series of solutions called the population (represented by chromosomes). The solution comes from the population and is used to create a new population. He was inspired by the hope that the new population would be better than the old one. Choose a solution to create a new solution based on its accuracy: the more efficient, the more likely it is to replicate. This process is repeated until certain conditions are met. The Genetic algorithm consists of the following steps:

1. [Start] Create the first random population of N chromosomes to get the best solution.
2. [Fitness] uses the fitness function $f(x)$ in the population to evaluate the fitness value of each x chromosome.
3. [New population] Follow these steps to create a new population before the end of the new population.
4. [Selection] Select the two parent chromosomes in the population based on their fitness values.
5. [Crossover] Create a new descendant on the parent's cross; otherwise, the descendants are an exact copy of the parent.
6. [Mutation] In each place, there is a new offspring/child mutation (position on the chromosome).
7. [Accept] Add new offspring/child to new population
8. [Replace] Select another newly created population for another step in the algorithm.
9. [Test] If the last condition is full, stop and return to the best solution in the current group.
10. [Loop] go to step 2.

13.7 Applications of Soft Computing

- Pattern Recognition
  - Visual
  - Sound
- Time-Series Forecasting
  - Financial
  - Weather
  - Engineering
- Diagnostic
  - Medicine
  - Engineering
- Robotics
  - Control
  - Navigation
  - Coordination
  - Object Recognition
- Process Control
  - Power Station
  - Vehicles or Missiles
- Optimization
  - Combinatorial problems like resource scheduling, routing etc.
- Signal Processing
- Speech & word Recognition
- Machine Vision
  - Inspection in manufacturing
  - Check Reader
- Financial Forecasting
  - Interest Rate
  - Stock Indices
  - Currencies
- Financial Services
  - Data Mining

  - Data Segmentation
  - Forecasting

- Handwriting Recognition
- Face Recognition
- Disease Recognition.

13.8 Future Scope of Soft Computing

The emergence of computer technology and its applications radically changed the aspects of science and industrial engineering. It gives scientists and engineers the opportunity to meet and overcome complex challenges related to the development of the system. The effective use of soft computing technologies will have a greater impact in the coming years. Soft computing technology played an important role in science and engineering. The field of soft computing is just beginning and there is a very fertile field of scientific research and engineering.

13.9 Conclusion

Today's growing technology demands require extremely complex systems that require sophisticated intelligent systems to ensure high performance under extreme conditions. The use of soft computing technology has made it possible to build complex systems that cannot support traditional systems because they lack accurate knowledge. In addition, smart systems are recognized areas in the field of intelligent systems.

This paper aims to raise public awareness about soft computing and bring traditional systems closer to intelligent systems. This paper also describes the evolution of the approach to intelligent technologies: soft computing, its applications and future prospects. Current paper scientists can better understand soft computing and its technology.

References

1. Zadeh, L.A.: Soft computing and fuzzy logic. IEEE J. Softw. **11**(6), 48–56 (1994)
2. Roy, S., Chakraborty, U.: Soft computing: neuro-fuzzy and genetic algorithms, 1st edn. Pearson Education (2013)
3. Deepa, S.N., Sivanandam, S.N.: Principles of Soft Computing, 2nd edn. Wiley (2011)
4. Haykin, S.O.: Neural Networks: A Comprehensive Foundation, 2nd edn. Prentice Hall (1998)
5. Kumar, P., Dixit, S., Singh, S.K.: Performance of aspect-oriented software quality modelling using artificial neural network technique. Int. J. Comput. Appl. **182**(36), 6–10 (2019)

6. Zadeh, L.A.: Fuzzy sets. Inf. Control **8**(3), 338–353 (1965)
7. Ross, T.J.: Fuzzy Logic with Engineering Applications, 3rd edn. Wiley (2011)
8. Kumar, P., Singh, S.K.: A framework for assessing the evolvability characteristics along with sub-characteristics in AOSQ model using fuzzy logic tool. In: IEEE International Conference on Computing, Communication and Automation (ICCCA-2017), pp. 340–345, Greater Noida, (Uttar Pradesh), India (2017)
9. Kumar, P., Singh, S.K.: Defect prediction model for AOP-based software development using hybrid fuzzy C-means with genetic algorithm and K-nearest neighbors classifier. Int. J. Appl. Inf. Syst. **11**(2), 26–30 (2016)
10. Goldberg, D.E.: Genetic Algorithms in Search, Optimization and Machine Learning. Addison Wesley (1989)

Chapter 14
Review of Different PAPR Reduction Techniques in FBMC-OQAM System

Ankita Agarwal and Rohit Sharma

Abstract Multicarrier modulation (MCM) schemes provide resistance against fading in wireless environment, due to which these techniques have been extensively worked upon for use in 4G and 5G wireless communication. Filter Bank Multicarrier is among the desired candidate for physical layer implementation in 5G communications. Irrespective of several features it suffers from power inefficiency due to high peak to average power ratio (PAPR). In this research study, different PAPR reduction techniques for Filter Band Multi-Carrier with Offset Quadrature Amplitude symbol mapping called as FBMC-OQAM are presented, such as non-linear Companding Transform, improved hybrid techniques based on conventional partial transmit sequence (PTS), iterative clipping and filtering (ICF), and Tone Reservation method.

Keywords FBMC/OQAM · PAPR reduction · Hybrid

14.1 Introduction

In 4G wireless Orthogonal Frequency Division Multiplexing (OFDM) which is one of multicarrier modulation (MCM) techniques were used in physical layer transmission as they can efficiently cope with frequency selective channels. But due to stringent requirement in 5G such as high spectral efficiency, less out of band radiation, 100% connectivity, availability, use of carrier aggregation etc. OFDM networks are not desirable [1]. For instance, the application of OFDM in the uplink of multiuser networks requires having full synchronization of the users' signals at the base station input which is difficult to maintain in mobile environments. Also the use of cyclic prefixes (CPs) and rectangular pulses on each subcarrier in OFDM introduces reduction of spectral efficiency and undesirable for filling spectrum holes in cognitive radio [2].

Filter Bank Multicarrier with Offset Quadrature Amplitude Modulation System (FBMC/OQAM) has attracted researchers for suitability particularly in multiple

A. Agarwal (✉) · R. Sharma
SRM Institute of Science and Technology, Ghaziabad, India
e-mail: kgec.ankita@gmail.com

V. E. Balas et al. (eds.), *Internet of Things and Big Data Applications*, Intelligent Systems Reference Library 180, https://doi.org/10.1007/978-3-030-39119-5_14

Table 14.1 Proposed PAPR reduction techniques

Notable techniques	Year	Reference
Tone reservation (TR)	2007	[6]
Partial transmit sequences (PTS)	2010	[7]
Selected mapping (SLM)	2010	[8]
Non-linear companding transform	2019	[9]
Improved bilayer partial transmit sequence and iterative clipping and filtering (IBPTS-ICF)	2018	[10]
Hybrid scheme based on Multi Data Block PTS and TR	2018	[11]
Segment based optimization (SBO)	2018	[12]

access and cognitive radio networks and 5G networks [3, 4]. FBMC/OQAM discard use of CP, higher spectral efficiency, low spectral leakage due to well localized filter design and suitable for emerging area of massive MIMO [2]. But common to the OFDM systems, issues related to the design of FBMC/OQAM systems is the high Peak-to-Average Power Ratio (PAPR) [5]. The high PAPR may result in significant distortion when such a signal enters non-linear region of high power amplifier which may limit its use in battery operated mobile applications. Hence, effective PAPR reduction techniques which are compatible with system architecture need to be studied. Some of the conventional techniques for reducing PAPR of OFDM and the modified hybrid techniques considering the overlapping structure of FBMC are listed in Table 14.1.

14.2 FBMC System Architecture

The architecture of FBMC-OQAM transmitter, which consists of N subcarriers, is presented in Fig. 14.1. Here the generated data undergoes three main processing steps namely OQAM pre-processing, IFFT, and Poly Phase filtering.

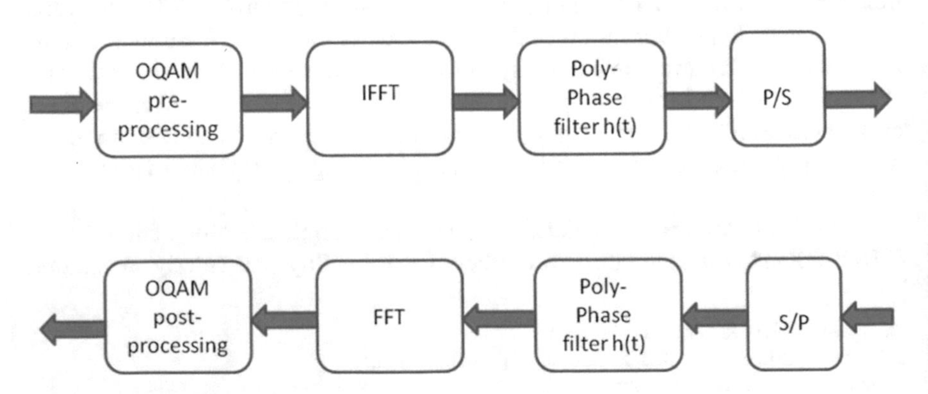

Fig. 14.1 FBMC-OQAM transmitter and receiver block diagram

The original data is QAM modulated and then complex QAM symbol S_m is serial to parallel converted. In FBMC offset QAM is used instead of conventional QAM so, to get OQAM symbol the real and imaginary parts are separated.

$$S_m^n = a_m^n + jb_m^n \tag{14.1}$$

Here, S_m is set of N complex input symbols $S_m^n, n = 0, 1, \ldots N-1$ and $m = 0, 1, \ldots M-1$. Here, N represents number of subcarrier and M denotes number of data blocks, 'a' represents the real part of OQAM symbols S_m^n which is staggered from 'b' its imaginary parts with a time offset of T/2. Here T is the symbol period. To generate Inphase (I) and quadrature (Q) valued input signal with π/2 phase difference, the symbols are multiplied by

$$\theta_{k,n} = j^{(k+n)} \tag{14.2}$$

So, in general for m-th incoming symbol, let $S_n^I(t) = Inphase\ input\ signal$ and $S_n^Q(t) = Quadrature\ input\ signal$. The I and Q signals are then transmitted on individual subcarrier where specially designed prototype filter h(t) is applied to each of them. In most literatures PHYDAS prototype [14] filter is considered. The m-th data block signal in time domain can be expressed as

$$S_m(t) = \sum_n = 1^N\{\{a_m^n\}h(t - mT) + j\{b_m^n\}h(t - mT - T/2)\}e^{j\Phi_{m,n}} \tag{14.3}$$

$\varphi_{m,n}$ is phase term = n(2πt/T + π/2). The transmitter realizations of FBMC with offset QAM is shown in Fig. 14.2. Transmitter consists of N parallel shaping filters for each I and Q component, which are frequency and time shifted versions of the same real low-pass prototype filter $h(t)$. Reverse operations are done at receiver.

The first specificity of this filter bank is the introduction of a shift of *T*/2 (symbol half-period) between the in-phase and in-quadrature components to mitigate the intersymbol interference. $h(t)$ is the prototype filter which extends over many symbol periods (*KT*) meaning that there is a time overlap between symbols, which significantly improves the data rate. In this paper, we consider a special and advantageous waveform referred to as the PHYDYAS prototype filter proposed in [14].

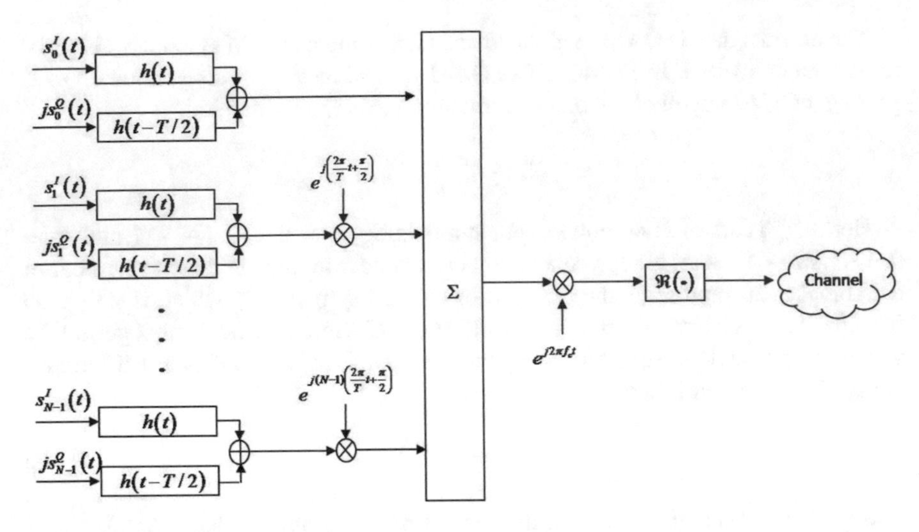

Fig. 14.2 FBMC transmitter design

14.3 The Peak to Average Power Ratio (PAPR) Calculation

The general continuous time equation of FBMC-OQAM system, which is summation of independent signals which are modulated on N subcarriers, results in signals with high fluctuation in amplitude from its average value. PAPR is defined as,

$$PAPR = PeakPower/AveragePower \tag{14.4}$$

Generally PAPR is expressed in dB as

$$\text{PAPR(dB)} = 10\log_{10} PAPR \tag{14.5}$$

To measure the efficiency and effectiveness of PAPR reduction technique parameter known as complementary cumulative distribution function (CCDF) is defined which denotes the probability of PAPR exceeding some threshold χ_0.

$$CCDF = Prob(PAPR \succ \chi_0) \tag{14.6}$$

Let $s[k]$ represents the discrete-time signal representation of continuous-time signal s(t), therefore to better approximate PAPR of $s[k]$ to its continuous time counterpart, FBMC/OQAM signal s(t) *is* β times oversampled. The oversampling factor β is generally taken to be greater than or equal to 4 [13].

$$PAPR(\mathrm{S}_k) = \frac{\max\{|S_k|^2\}}{E\{|S_k|^2\}} \tag{14.7}$$

14.4 Techniques for PAPR Reduction

In this section some of the PAPR reduction schemes compatible with FBMC overlapping structure are presented.

14.4.1 Non-linear Companding Transform Techniques

Companding transform represents the signal with large amplitude fluctuation in its envelope into a uniformly distributed signal at transmitter and equivalent inverse companding at receiver. This technique is less complex and has good Bit Error Rate (BER) as the process requires increases in the smaller signal power levels at the transmitter. The major drawback of uniform linear companding is that the transmitter has to transmit extra power than before as average power value is enhanced. Therefore in [9] Non-linear companding is applied at transmitter used in which small amplitude signal are enlarged and high signal values are compressed using strict monotone function. In this technique, the output of polyphase filter S_K undergo companding to generate companded signal v_k

$$v_k = h(s_k) \tag{14.8}$$

where h(.) is companding function. In [9] author has presented different companding function summarized below with their transmitter function as,

Hyperbolic tangent companding transmitter equation is given by

$$h(S_k) = \alpha \tanh(\beta S_k) \tag{14.9}$$

Error function companding: This companding equation is expressed as

$$h(s_k) = \beta erf(\beta S_k) \tag{14.10}$$

Logarithm function companding is expressed as

$$h(s_k) = \alpha \log_e(1 + \beta S_k) \tag{14.11}$$

In above equations, parameters α and β are positive values defining the level of Companding.

Mu-law companding technique is expressed as

$$h(S_k) = u(Mu)\text{sgn}(S_k)\frac{\ln(1 + Mu|S_k|)}{\ln(1 + Mu)} \tag{14.12}$$

here the Mu ratio controls the level of companding at transmitter, where $u(Mu)$ is constant of normalization.

A-law companding is given as

$$h(S_k) = k(A)\mathrm{sgn}(S_k)\begin{cases} \frac{A|S_k|}{1+\ln A}, & if\ |s_k| \prec |s_k|\max/A \\ \frac{1+\ln(A|S_k|)}{1+\ln A}, & if\ |S_k| \geq |S_k|\max/A \end{cases} \quad (14.13)$$

where $sgn(y)$ = sign of the input, $|S_k|$ = magnitude value of sk and the value A controls the level of Companding.

14.4.2 Improved Bilayer Partial Transmit Sequence and Iterative Clipping and Filtering Scheme (IBPTS-ICF)

In [10] author has discussed a joint scheme for PAPR reduction combining linear and a non-linear scheme namely Partial transmit sequence (PTS) and iterative clipping and filtering (ICF). The proposed method first reduces the probability of peak values by IBPTS method and then clipping and filtering the processed signal using ICF.

Conventional PTS scheme used for OFDM divides the data block S_m into V disjoint sub blocks termed as S_v, where v = 1,2, … V. Then optimal phase factors are searched from a given set to independently rotate the sub-carriers in all sub-block by these phase values. Here search complexity is reduced by limiting the number of phase factors.

Let W = number of allowed phase factors, then vector **b** of these factors is

$$b = \{b^v = e^{j2\pi l/W}, l = 0, 1, \ldots W - 1\} \quad (14.14)$$

For OFDM the phase rotation method is applied only on current data block but because FBMC has overlapping structure this scheme gives degraded result. In FBMC/OQAM m-th block of data coincides with next (β − 1) data sub blocks.

Conventional clipping scheme reduce PAPR by simply limiting the peak signal value of discrete time FBMC/OQAM sample s[k] of m-th incoming symbol to some threshold value A_{max}.

$$A_{max} = \gamma.\sigma \quad (14.15)$$

Here, γ is clipping Ratio and σ is root mean square power of the FBMC signal. Clipping introduces non-linear effects and in-band signal distortion can occur, giving high BER which reduce spectral efficiency.

IBPTS-ICF joint optimization scheme uses IBPTS and ICF schemes which are modified forms of above two mentioned conventional schemes. In IBPTS scheme a bilayer structure is proposed and numbers of phase factors are kept finite to decrease complexity. Bilayer structure consist of first layer 1 which is Frame of M overlapping data blocks denoted by $\boldsymbol{S}_m$, m = 0, 1, … M − 1 and then each m-th data block is partitioned into V sub block constituting layer 2.

$$S_m = \sum_{v=1}^{V} S_m^v \tag{14.16}$$

Optimal phase factor now is $\mathbf{b_m} = \left[\tilde{\dot{b}}_m^1, \ldots \tilde{\dot{b}}_m^v \ldots \tilde{\dot{b}}_m^V \right]$. In this scheme to further reduce computational complexity penalty factors are introduced. Penalty factor is a special threshold, which is used for decreasing the required number of data blocks needed to be searched, hence improving the complexity of computations greatly. In proposed algorithm phase factors $\tilde{\dot{b}}_m^1$ can be {−1, 1} with W = 2. These penalty factors are applied for both layers denoted by ω_p and μ_p.

After IBPTS scheme, to further decrease PAPR, ICF algorithm is used in which the signal is first clipped and then filtered to reduce out of band radiation. However still in-band distortion exist which is resolved by iterative compensation strategy described in [10].

14.4.3 Hybrid Multi Block PTS and TR Method

Two other hybrid approaches based on conventional PTS and tone reserve TR which are both probabilistic PAPR reduction methods proposed by author in [11] called as hybrid PTS-TR scheme and multi hybrid (M-hybrid) PTS-TS scheme. The first scheme considers overlapping effect in only two adjacent data blocks and the second scheme also takes into account multi data blocks overlapping in each segment signal.

For the proposed technique, s(t) is segmented into M + β intervals with interval duration T and conventional PTS and TR schemes modified according to FBMC structures is used. Conventional PTS scheme is already described.

Conventional Tone Reservation scheme allows some subcarriers which are either unused or are reserved especially to accommodate a peak-cancelling signal that brings down the PAPR of the time domain signal. Out of the total N subcarriers also called as tones, R subcarriers which are usually null subcarriers are used as peak reduction tones (PRT) and remaining N-R are data tones. The m-th data block S_m^n is bifurcated into data vector (D_m^n) and PAPR reduction vector (C_m^n).

$$S_m^n = D_m^n + C_m^n = \begin{cases} C_m^n, & n \in R \\ D_m^n, & n \in R^C \end{cases} \tag{14.17}$$

Thus, optimised peak reduction signal $c(t)$ is added to FBMC signal $s(t)$ in time domain to reduce peak power to some limited range decreasing overall PAPR.

Hybrid PTS-TR scheme divides M incoming data blocks into multiple segments depending on overlapping factor i.e. M/β and each segment has β blocks. The first data block in the first segment is divided into V disjoint subblocks and are optimized using PTS approach described above. In hybrid scheme second data block is optimised by finding the optimal phase rotation factors which are generated by selecting the minimum power between the first and the second data blocks. Similarly all optimized segment signals are added to obtain final PAPR reduced signal. In M-hybrid scheme, a data block is optimized by minimizing signal power between current data block and previous multi data blocks.

14.5 Conclusion

In this paper, few PAPR reduction schemes available in literature applicable for emerging FBMC OQAM technique are discussed. These includes conventional and hybrid techniques with their advantages and drawbacks. The summary of these recent works is given in Tables 14.1 and 14.2 in which the proposed technique is stated to be the best one in yielding the expected results.

Table 14.2 PAPR at CCDF = 10^{-3} of notable PAPR reduction techniques

Reference	Reduction technique	Modulation format and companding factor	Prototype filter	PAPR (dB)
[9]	Conventional FBMC	512 subcarriers, 128-OQAM	Square Root Raised Cosine filter, overlapping factor = 6 and Roll-off factor = 0.53	18.41
	Mu-law	512 subcarriers, 128-OQAM, Mu = 255		6.629
	A-law	512 subcarriers, 128-OQAM, A = 87.7		7.879
[10]	Conventional FBMC	128 subcarriers, 4-OQAM	PHYDAS	12
	IBPTS-ICF	M = 4, β = 4, V = 4, ω_P = 2, γ = 1.2, No. of iteration after clipping = 5		3.8
[11]	Conventional FBMC	64 subcarriers, 4-OQAM	Square Root Raised Cosine filter, length of filter = 4T Roll-off factor = 1	10.1
	Hybrid scheme	64 subcarriers, 4-OQAM, β = 4, *W* = 2, V = 4, c = 8, A = 2.4		7.3
	M-hybrid scheme	64 subcarriers, 4-OQAM, β = 4, *W* = 2, V = 4, c = 8, A = 2.4		7.1

References

1. Banelli, P.: Modulation formats and waveforms for 5G networks: who will be the heir of OFDM? An overview of alternative modulation schemes for improved spectral efficiency. IEEE Signal Process. Mag. **31**(6), 8–93 (2014)
2. Ihalainen, T.: Channel equalization for multi-antenna FBMC/OQAM receivers. Trans. Veh. Technol. **60**(5), 2070–2085 (2011)
3. Farhang-Boroujeny, B.: OFDM versus filter bank multicarrier. IEEE Signal Process. Mag. **28**(3), 92–112 (2011)
4. Farhang-Boroujeny, B.: Cosine modulated and offset QAM filter bank multicarrier techniques: a continuous-time prospect. EURASIP J. Advant. Signal. Process. **20**, 1–16 (2010)
5. Schellmann, M., et al.: FBMC-based air interface for 5G mobile: challenges and proposed solutions. In: Proceedings of 9th International Conference on Cognition Radio Oriented Wireless Networks and Communications (CROWNCOM), pp. 102–107 (2014)
6. Devlin, C.A., Zhu, A.: Gaussian pulse based tone reservation for reducing PAPR of OFDM signals. In: Proceedings of the IEEE 65th Vehicular Technology Conference, VTC, pp. 3096–3100 (2007)
7. Ku, S. J.: A reduced-complexity PTS based PAPR reduction scheme for OFDM systems. IEEE Trans. Wireless Commun. **9**(8), 2455–2460 (2010)
8. Li, C.-P.: Novel low-complexity SLM schemes for PAPR reduction in OFDM systems. IEEE Trans. Signal Process. **58**(5), 2916–2921 (May 2010)
9. Imad, A.: Performance evaluation of PAPR reduction in FBMC system using nonlinear companding transform. ScienceDirect ICT Express **5**, 41–46 (2019)
10. Zhao, J., et al.: PAPR reduction of FBMC/OQAM signal using a joint optimization scheme. Digital Object Identifier: https://doi.org/10.1109/access.2017.2700078
11. Wang, H., et al.: Hybrid PAPR reduction scheme for FBMC/OQAM systems based on multi data block PTS and TR methods. IEEE Access. **4**, 4761–4768 (September 2016)
12. Moon J.-H. et al.: PAPR reduction in the FBMC-OQAM system via segment-based optimization. IEEE Access. **6**, 4994–5002 (2018)
13. Shaheen, I.A.: Proposed new schemes to reduce PAPR for STBC MIMO FBMC systems. Commun. Appl. Electron. **6**(9), 27–33 (2017)
14. Bellanger, M., et al.: FBMC physical layer: a primer, PHYDYAS FP7Project Document (2010)

Chapter 15
A Comprehensive Study for Security Mechanisms in Healthcare Information Systems Using Internet of Things

Y. Harold Robinson, R. Santhana Krishnan and S. Raja

Abstract For developing the cyber physical based smart pervasive architecture, Internet of Things (IoT) plays a vital role. The IoT provides a huge amount of applications like healthcare system. IoT is useful for constructing the latest healthcare developments which supports the technology-based, economic-based and social-based application. This chapter provides the detailed study of the emerging technology based on the IoT based healthcare mechanism with security issues. Additionally, this chapter distinguishes the IoT based security mechanism for providing the security which also diminishes the healthcare facilities. The collaboration model minimizes the risk for implementing the e-Healthcare policies all over the world to decide the economical and social solutions for the Healthcare Information Systems.

Keywords Internet of Things · Health information · Exchange · Health information system · Utilization

15.1 Introduction

The Evolution of State Health Information Exchange stated that the utilization of HIE has numerous important result in dissimilar project designs such as financing, patterns of success, programmatic sustainability and challenges to recognize the trends and superlative practices [1]. The occasion are driven by assuming HIE to progress the excellence and minimize the expenditure of health care and to increase the workflow of quantifiable and management information among health care system

Y. Harold Robinson (✉)
School of Information Technology and Engineering, Vellore Institute of Technology, Vellore, India
e-mail: yhrobinphd@gmail.com

R. Santhana Krishnan
Department of Electronics and Communication Engineering, SCAD College of Engineering and Technology, Cheranmadevi, India
e-mail: santhanakrishnan86@gmail.com

S. Raja
Department of Mathematics, SCAD College of Engineering and Technology, Cheranmadevi, India
e-mail: nellairajaa@yahoo.com

V. E. Balas et al. (eds.), *Internet of Things and Big Data Applications*, Intelligent Systems Reference Library 180, https://doi.org/10.1007/978-3-030-39119-5_15

[2]. Furthermore, data exchange within the healthcare systems is amongst the most complex issues of electronic Health Information Exchange in health systems and is moreover, considered as one of the most demanding troubles in electronic health records management [3]. If the data is not obtainable for numerous users like planners, managers, policy makers, individuals, communities and healthcare providers, the data would have modest significance [4]. Additionally, the health information system is incapable to present an attentive for sustain health patient position and untimely warning potential [5]. Consequently, allocation and communication using smart technology are significant characteristics for the health information system.

HIE has the prospective to engage collaboration and exchanges data within the healthcare providers and stakeholders with patients [6]. There is a require to replace health information by utilizing IoT services to increase the eminence and security of healthcare by properly, resourcefully and steadily utilize a patient's medical records and remote monitoring information that are accessible [7]. The ability to exchange data or mechanism within several systems is called as the interoperability framework [8]. Access to the interoperability of healthcare in the public sector is usually related to capacity of services to present a framework for information exchange within the dissimilar users, measures, and process policy makers [9].

15.2 Related Works

HIE is certainly one of the interior constituents of e-Health [10]. It may assist a medical organization in making superior and added suitable medical implications by exploiting the latest technology to broadcast in real-time, patient's health care data to any medical organization that deserves it [11]. Precedent research has been implemented on how the HIE methodology may assist to elevate the superiority of medical care, increase patient protection and diminish medical expenditures [12]. When patients travel from one hospital to another, their medical details are reserved in numerous papers and electronically-related devices consequently; there is no research directory to position their medical details [13]. This concealed personality of patient's medical details may escort to supplementary procedures, replacement tests and several problems including unfavourable medicine communication. According to the related studies, the output have indicated and explained the reimbursement of HIE and this is why numerous healthcare contributors discover the system complicated to appreciate and employ [14]. Most healthcare officials acquire the electronic patient data from outside their practice assets such as drug data management and labs [15]. There are a number of difficulties facing the exploit of HIE like insufficient visibility and information as to where patient's medical details may be pursued. There is also, the deficient of convenience and information standards which construct it complicated to transfer clinical data and complex devices [16]. Also, there are various non-technology barriers such as trouble burden problems of patient permission, differences of business representation, restricted appreciative, defeat of reasonable improvement [8].

Consistent with, dissimilar studies in some countries, HIS has been reported to be opposite a lot of difficulties and issues. There are 2 types of HIE: public HIE and private HIE [17]. The value of exchange data does not ensure medical value for exchanged information because it is just an obligatory obligation and most of them do not use auto-query [18]. Other studies have reported the imperfect research on present health IT to provide the necessary support of health collaboration [19]. Another study highlighted the issue of nonflexible and inadequate information records to congregate the requirements for practice, thus, limiting access to health assets and practice [20]. The user needs affirmative approval to patient information access but no such approval is needed to construct information available on regional HIE [21]. In an urgent situation, the user can utilize the patient information without affirmative permission. Therefore, the execution of HIE is not enough to conquer these real-time issues [22].

Another 2 studies accounted that there is a necessary for corrective measures to increase the competence of actual exploit and sustainability of the health system [23]. The healthcare officials have organized to absolute the building of health exchange during future but there is still a necessary to increase the practice competence as well as sustainability of finished systems [24]. The lack of shared decision-making and untrustworthy health information for objective patients produces it hard to appreciate and converse within dissimilar health information systems that influence analysis and treatment methods [25].

The HIE is not extensively adopted in several health care managements [26]. Additionally, user's face numerous difficulties such as limited convenience to health patient records and the requirement that distinguish and completely embrace the latest techniques such as IoT and cloud computing usages [27]. Further investigation in several occupational personnel in the health sector recommended that there was extensive use of electronic health details but there was modest attention in using HIE [28]. In conversation with the nurses in the hospitals, it was initiated that there was a necessary for widespread message and data sharing within the health care employees in related organizations to assist flat transitions and healthcare data technologies are introduced to ease these developments [29].

There are fewer infrastructures for the extent of hospital management and confrontation to using the application of latest technology in addition to a lack of planning for information recovery [30]. Because of that there is modest use of HIS and no important correlation within the HISs that exists. Scientists have planned to solve this issue by integrating the needs and using the latest technology to gather data and improve human resource management and access to services [31]. At last, after reviewing numerous studies conducted and analyzing the confront they features it can be seen that there is a requirement to exploit latest development technologies like IoT in the healthcare segment and particularly in the HIE system to sufficiently address the current problems.

15.3 Methodology

The Health data within the providers for the main and secondary information systems are also having the main problems. The patient record could not maintained properly because of the unqualified staff, so we need to find the qualified and well trained staff to improve the workflow and avoid the unfriendliness of the user using the IoT technique. Some part-time clinicians enlarge coordination issues because of using the low employ with HIE technologies in their office administration. Vendor communities need for superior collaboration and integration efforts Techniques for reducing error and increasing the training structure in their organization. It is very difficult to understand the user's needs and very difficult to acquire significant historical patient data, which is including the medication data, test results, and medical histories (Fig. 15.1).

Communication and sharing data within the General Practitioners (GPs) does not work well to the relevant organization. The HIE it's not widespread adopted for the purpose of communication. The limited in accessibility is needed for health patient record and data. The adoption of an HIE needs to full realization of the new technology benefits Employee's resistance. The system may have the low capability of storing doctor consultations. It also lacks interoperability standards and implementation guidelines lacks interoperability and collaboration between kinds of hospitals. Lacks government roles and understanding for making the supportive environment in the global is the common mistake. Lacks agreement about what EHRs abilities mean and constitute. Implementation and adoption of EHR does not achieve the desired rate of distribution Poor hosp ital management infrastructure. Resistance in using IT

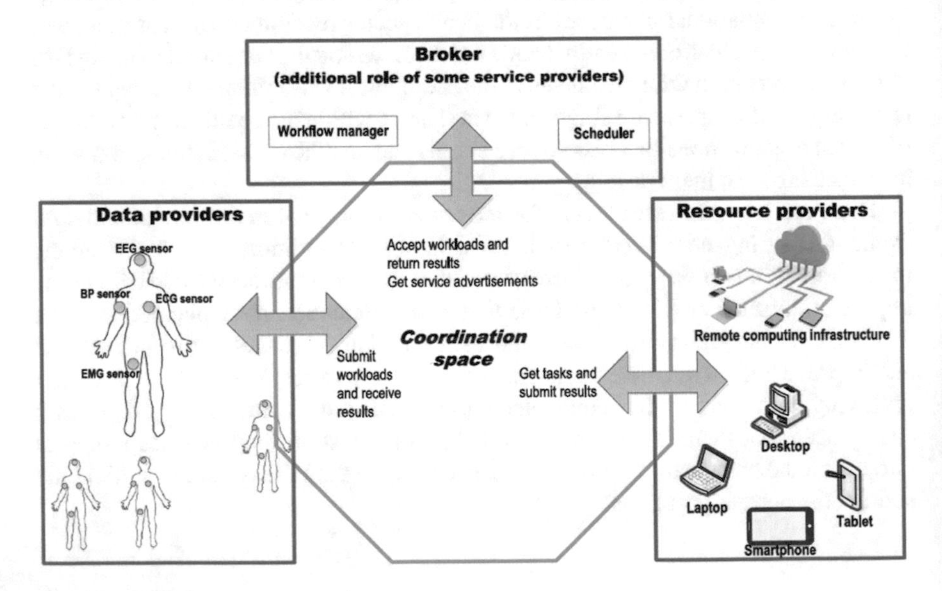

Fig. 15.1 IoT based healthcare information system

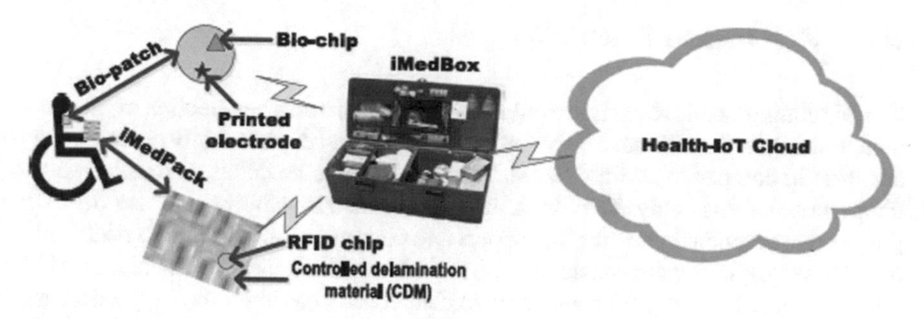

Fig. 15.2 Intelligent health information system

application is very less and lack of planning and data recovery. Figure 15.2 demonstrates the Intelligent Health Information system which consists of Health based IoT and Cloud system with the patient data is sharing using the electrode, bio chip and RFID chip and also the controller for communication.

Moreover, the IoT-based healthcare system has some disadvantages that make its adoption a concern, shown by the summary below. The IoT network is both diverse and complex. Multiple services are needed to produce device counts; massive increases in internet bandwidth are necessary to drive requirements for lower latency, better determinism and processing closer to the perimeter of the network. Thus, any failure or bugs in the software or hardware will consequence in solemn consequences. A single power breakdown for example, can cause a lot of problem. The compatibility difficulty between the heterogeneous components will allow several problems in the system. When the products of dissimilar manufacturers are interconnected, the problem of compatibility arises because there is no common standard agreed to among the different manufacturers. To make your mind up this incompatibility problem, people will be forced to buy from a single manufacturer rather than from different manufacturers, thus, creating domination for one manufacturer. This stops users from shopping for the preeminent obtainable products from various manufacturers in the market. Security and privacy identifies the location of persons or objects in real time and gathers inappropriate personal record, thus creating more challenges in a dynamic manner. The problem of protecting identities and privacy would arise and it can also be seen that the enormous information from millions of things in a healthcare system could lead to many security challenges.

Subsequently it is important to handle and sustain the HIE to healthcare among the organizations. However, to guarantee effective usage of IoT in the healthcare environment, there must be careful focus on a number of factors with different perspectives that may hold technical and system factors of information technology and attributes of the organization which present the technology as well as the reaction of individuals in the organization toward the use of innovative technology. Consequently, study presents an introductory study to recognize the key factors related to the utilization of IoT services to promote the current system of HIE.

15.4 Results and Discussion

The population of this study is the public healthcare sector. The selection of this population was based on the necessary for functionality and dependability of technology adoption in comparison with other statutes done. Some hospitals were selected for the purpose of this study. They were chosen for their capability and also for their prospective potential in healthcare support and concentration of technological facilities. These hospitals were considered the main hospitals in the implementation of IT infrastructure. The sample hospitals were also chosen based on the suggestion and assistance of the Ministry of Health and taking into account their number of patients, beds, physicians and technicians.

The physicians reported a number of difficulties in their decision-making because of poor patient's health data and some dissimilarity in receiving treatment when the patients visited special physicians. The physicians expressed some concerns about the dependability of data with records and the occurrence of medical errors. Some cases needed to be monitored for a long period and previous details had to be checked-if they were obtainable. Most of the physicians admitted using social media to share the data about some cases among multi-specialist doctors but that necessary more time to explain and to upload the results. These physicians recommended using the HIE to assist them in sharing the Patient ID with each other in order to be proficient to observe the patient's health position in real-time and anywhere. The Interviewees from the Ministry of Health reported; "if we can enable the electronic HIE system to be used among hospitals it will help the physicians and decision makers to present better healthcare and support the Ministry of Health with weekly reports and inform them of some disease outbreak". The Ministry of Health needs to ascertain an innovative healthcare system utilizing novel technology amongst different healthcare providers.

The physicians and IT staff reported that training is an important factor to encourage frequent use of the system not only during the first stages of the system set up and implementation but also, during training. Proper training can escort to improve the deployment of the system use. Cooperation within the healthcare staff by filling up the data and maintaining the reports of patient position would provide a clear patient data record and make it easily understandable for other users.

The reason for the nonexistence of an electronic health record exchange system within the hospitals could be the desire of personal hospitals to maintain the data to them. Also, they highlighted some other issues such as network capability in their respective organizations as well as concerns about security and privacy for their patients and low connection that made it difficult to access the patient's data. Most of the HISs in the health sector do not use cloud computing, RFID and IoT. Also, unfortunately, the health data is saved nearby and not communicated with any other patient systems. Hence, the workflow, cost-effectiveness, training and cooperation play as key factors to manipulate the organizational domain to use IoT technology in health data exchange system and it is illustrated in Fig. 15.3.

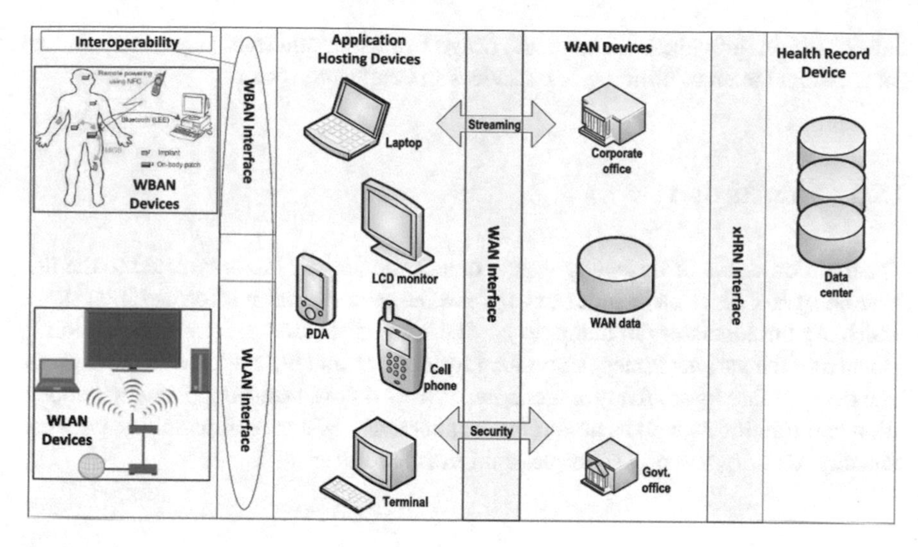

Fig. 15.3 Interoperability in IoT healthcare information systems

The entire adoption of the IoT in the field of healthcare sector is critical to measure and determine that the individual features of IoT security requirements. The threat related models and measurements from the healthcare are illustrated in Fig. 15.4.

In addition, other factors related to the technological domain in terms of network capability, security and privacy, compatibility and ubiquitous connectivity. It is observed that the other system factors are imposed in terms of accessibility, usefulness factors which have a significant effect on individuals of utilizing IoT services in the health information exchange. Finally, there are multiple factors that formed

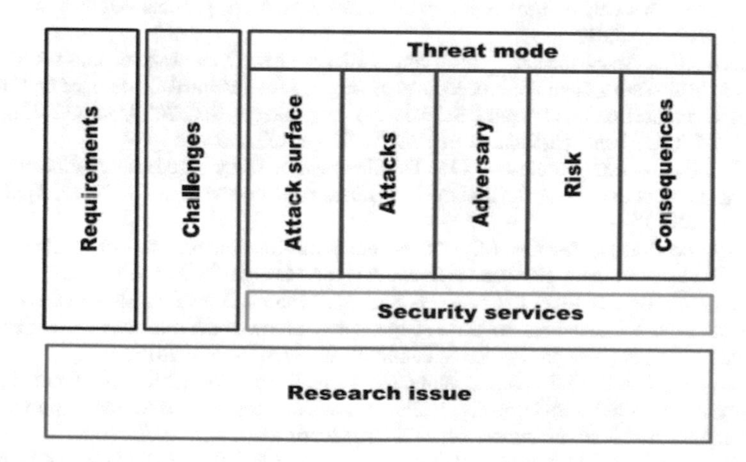

Fig. 15.4 Issues in IoT security mechanism

have effect on individual's. The actual usage behaviour and trust play as key factors for a medical team to utilizing IoT services in healthcare sectors.

15.5 Conclusion

The main objective of this study was to determine the key factors related to the utilization of IoT services in order to improve the current health information systems which are unsatisfactory in many ways. The study discussed the current situation in global and the various issues facing the introduction and utilization of IoT services. The relevant factors are many and encompass: workflow, cost-effectiveness, cooperation and training as well as ubiquitous connectivity, system compatibility, network capacity, security and privacy of patient information.

References

1. Al-Aswad, A.M., Brownsell, S., Palmer, R., Nichol, J.P.: A review paper of the current status of electronic health records adoption worldwide: The gap between developed and developing countries. J. Health Inf Dev. Countries **7**, 153–164 (2013)
2. Ahrnadian, L., Nejad, S.S., Khajouei, R.: Evaluation methods used on Health Information Systems (HISs) in Iran and the effects of HISs on Iranian healthcare: A systematic review. Int. J. Med. Inf. **84**, 444–453 (2015)
3. Krishnan, R.S., Julie, E.G., Robinson, Y.H., Kumar, R., Son, L.H., Tuan, T.A., Long, H.V.: Modified zone based intrusion detection system for security enhancement in mobile ad-hoc networks. Wireless Netw. 1–15 (2019)
4. Campion, T.R., Edwards, A.M., Johnson, S.B., Kaushal, R.: Health information exchange system usage patterns in three countries: Practice sites, users, patients and data. Int. J. Med. Inf. **82**, 810–820 (2013)
5. Darshan, K.R., Anandabunar, K.R.: A comprehensive review on usage of Internet of Things (IoT) in healthcare system. In: Proceedings of the 2015 International Conference on Emerging Research in Electronics, Computer Science and Technology (ICERECT), pp. 132–136, 17–19 Dec 2015. IEEE, Mandya, India (2015). ISBN: 978-1-4673-9563-2
6. Balaji, S., Julie, E.G., Robinson, Y.H.: Development of fuzzy based energy efficient cluster routing protocol to increase the lifetime of wireless sensor networks. Mob. Netw. Appl. **24**(2), 394–406 (2019)
7. Dobrzykowski, D.D., Tarafdar, M.: Understanding information exchange in healthcare operations: Evidence from hospitals and patients. J. Oper. Manage. **36**, 201–214 (2015)
8. Kamradt, M., Baudendistel, I., Langs, G., Kiel, M., Eckrich, F., et al.: Collaboration and communication in colorectal cancer care: A qualitative study of the challenges experienced by patients and health care professionals. Family Pract. **32**, 686–693 (2015)
9. Downing, N.L., Adler-Milstein, J., Palma, J.P., Lane, S., Eisenberg, M., et al.: Health information exchange policies of 11 diverse health systems and the associated impact on volume of exchange. J. Am. Med. Inf Assoc. **24**, 113–122 (2016)
10. Balaji, S., Julie, E.G., Robinson, Y.H., Kumar, R., Thong, P.H., Son, L.H.: Design of a security-aware routing scheme in mobile ad-hoc network using repeated game model. Comput. Stan. Interfaces **66** (2019)

11. Fernandez, F., Fallis, G.C.: Opportunities and challenges of the Internet of Things for healthcare: Systems engineering perspective. In: Proceedings of the 2014 EAI 4th International Conference on Wireless Mobile Communication and Healthcare (Mobihealth' 1 4), pp. 263–266, 3–5 Nov 2014. IEEE, Athens, Greece (2014). ISBN: 978-1-4799-5024-9
12. Robinson, Y.H., Balaji, S., Julie, E.G.: FPSOEE: Fuzzy-enabled particle swarm optimization-based energy-efficient algorithm in mobile ad-hoc networks, J Intell. Fuzzy Syst. **36**(4), 3541–3553 (2019)
13. Hameed, R.T., Mohamad, O.A., Hamid, O.T., Tapus, N.: Design of E-Healthcare management system based on cloud and service oriented architecture. Proceedings of the 2015 Conference on E-Health and Bioengineering (EHB' 1 5), pp. 1–4, 19–21 November 2015. IEEE, Iasi, Romania (2015). ISBN: 978-1-4673-7544-3
14. Hekrnat, S.N., Dehnavieh, R., Behmard, T., Khajehkazemi, R., Mehrolhassani, M.H., et al.: Evaluation of hospital information systems in Iran: A case study in the Kerman Province. Glob. J. Health Sci. **8**, 95–95 (2016)
15. Robinson, Y.H., Rajaram, M.: A memory aided broadcast mechanism with fuzzy classification on a device-to-device mobile ad hoc network **90**(2), 769–791 (2016)
16. Kadhurn, A.M., Hasan, M.K.: Assessing the determinants of cloud computing services for utilizing health information systems: A case study. Int. J. Adv. Sci. Eng. Inform. Technol. **7**, 503–510 (2017)
17. Robinson Y.H., Julie E.G., Saravanan, K., Kumar, R., Son, L.H.: DRP: Dynamic routing protocol in wireless sensor networks, wireless personal communications, pp. 1–17. Springer (2019)
18. Latif, A.I., Othman, M., Sulirnan, A., Daher, A.M.: Cwrent status, challenges and needs for pilgrim health record management sharing network, the case of Malaysia. Int. Arch. Med. **9**, 1–10 (2016)
19. Robinson, Y.H., Rajaram, M.: Energy-aware multipath routing scheme based on particle swarm optimization in mobile ad hoc networks. Sci. World J. 1–9 (2015)
20. Li, Y.C.J., Yen, J.C., Chiu, W.T., Jian, W.S., Syed-Abdul, S., et al.: Building a national electronic medical record exchange system—experiences in Taiwan. Comput. Methods Programs Biomed. **121**, 14–20 (2015)
21. Lim, A.K., Thuemmler, C.: Opportunities and challenges of internet-based health interventions in the future internet. In: Proceedings of the 12th International Conference on Information Technology-New Generations (JING' 15), pp. 567–573, 13–15 Apr 2015. IEEE, Las Vegas, Nevada (2015). ISBN: 978-1-4799-8827-3
22. Robinson, Y.H., Julie, E.G., Balaji, S., Ayyasamy A.: Energy aware clustering scheme in wireless sensor network using neuro-fuzzy approach. Wireless Pers. Commun. **95**(2), 703–721 (2017)
23. Mastebroek, M., Naaldenberg, J., van den Driessen Mareeuw, F.A., Leusink, G.L., Lagro-Janssen, A.L., et al.: Health information exchange for patients with intellectual disabilities: A general practice perspective. Br. J. Gen. Pract. **66**, e720–e728 (2016)
24. Balaji, S., Robinson, Y.H., Julie, E.G.: GBMS: A new centralized graph based mirror system approach to prevent evaders for data handling with arithmetic coding in wireless sensor networks, Ingénierie des Systèmes d'Information **24**(5), 481–490 (2016)
25. Mishuris, R.G., Yoder, J., Wilson, D., Mann, D.: Integrating data from an online diabetes prevention program into an electronic health record and clinical workflow, a design phase usability study. BMC. Med. Inf. Decis. Making **16**, 1–13 (2016)
26. Richardson, J.E., Ves, J.R., Green, C.M., Kem, L.M., Kaushal, R., et al.: A needs assessment of health information technology for improving care coordination in three leading patient-centered medical homes. J. Am. Med. Inf. Assoc. **22**, 815–820 (2015)
27. Robinson, Y.H., Julie, E.G.: MTPKM: Multipart trust based public key management technique to reduce security vulnerability in mobile ad-hoc networks, Wireless Pers. Commun. **109**, 739–760 (2019)
28. Thannalingam, S., Hagens, S., Zehner, J.: The value of connected health information: Perceptions of electronic health record users in Canada. BMC. Med. Inf. Decis. Making **16**, 93–100 (2016)

29. Vest, J.R., Kem, L.M., Campion Jr., T.R., Silver, M.D., Kaushal, R.: Association between use of a health information exchange system and hospital admissions. Appl. Clin. Inf. **5**, 219–231 (2014)
30. Robinson, Y.H., Krishnan, R.S., Julie, E.G., Kumar, R., Son, L.H., Thong, P.H.: Neighbor knowledge-based rebroadcast algorithm for minimizing the routing overhead in mobile ad-hoc networks. Ad Hoc Netw. **93**, 1–13 (2019)
31. Zdravkovic, M., Noran, O., Trajanovic, M.: On pervasive health information systems in the internet of things. In: Proceedings of the 25th Australasian Conference on Information Systems, pp. 1–10, 8–10 Dec 2014. Auckland University of Technology, Auckland, New Zealand (2014)

Chapter 16
Predictive Algorithm and Criteria to Perform Big Data Analytics

C. Gopala Krishnan, E. Golden Julie and Y. Harold Robinson

Abstract The internet is faster than the speed of light, memory storage and computing power has moved to the cloud. Big Data Analytics plays a vital role to segregate the data in some order. There is a huge amount of data available in the Information Industry. This data is of no use until it has converted into useful information. It is necessary to analyse this huge amount of data and extract useful information from it. An algorithm in data mining (or machine learning) is a set of heuristics and calculations that creates a model from data. To create a model, the algorithm first analyses the data you provide, looking for specific types of patterns or trends. The algorithm uses the results of this analysis over much iteration to find the optimal parameters for creating the mining model. These parameters have been applied across the entire data set to extract actionable patterns and detailed statistics. Included in this category is a very advanced technique and tool called "predictive algorithms". Predictive algorithms have revolutionized the way we view the future of data and have demonstrated the big strides of computing technology. In this paper, we discussed about the criteria used to choose the right predictive model algorithm.

Keywords Predictive analysis · Data processing · Future data · Fraud detection

---

C. Gopala Krishnan
Department of Computer Science and Engineering, Francis Xavier Engineering College, Tirunelveli, India
e-mail: skywarekrish@gmail.com

E. Golden Julie
Department of Computer Science and Engineering, Anna University Regional Campus, Tirunelveli, India
e-mail: goldenjuliephd@gmail.com

Y. Harold Robinson (✉)
School of Information Technology and Engineering, Vellore Institute of Technology, Vellore, India
e-mail: yhrobinphd@gmail.com

V. E. Balas et al. (eds.), *Internet of Things and Big Data Applications*, Intelligent Systems Reference Library 180, https://doi.org/10.1007/978-3-030-39119-5_16

16.1 Introduction

Descriptive analysis is used to describe the basic features of the data in the study. They provide simple summaries about the sample and the measures. Together with simple graphical analysis, they form the basic virtual of any quantitative analysis of data. With descriptive analysis, one simply describes what is or what the data shows. Description of data is needed to determine the normality of the distribution, description of the data is necessary as the nature of the techniques to be applied for inferential analysis of the data depends on the characteristics of the data. But first, let's break down the process of predictive analytics into its essential components. For the most part, it can be dissected into 4 areas.

16.1.1 Descriptive Analysis

It consists of systematic observation and description of the characteristics or properties of objects or events for the purpose of discovering relationships between variables. The ultimate purpose is to develop generalizations that may be used to explain phenomena and to predict future occurrences. To conduct research, principles must be established so that the observation and description have a commonly understood meaning. Measurement is the most precise and universally accepted process of description, assigning quantitative values to the properties of objects and events [1]. Planning and care in research design and data collection provides a substantial guarantee of quality in research but the ultimate test lies in the analysis [2]. Data in the real world often comes with a large quantum and in a variety of formats that any meaningful interpretation of data cannot be achieved straightway [3]. In order to achieve the objectives of the study, analysis of the data collected forms an important and integral part. Analysis means categorizing, classifying and summarizing data to obtain answers to the research questions. Classification also helps to reduce the vast data into intelligible and interpretable forms [4].

Descriptive analysis of data limits generalization to a particular group of individuals observed. No conclusions extend beyond this group and any similarity to those outside the group cannot be assumed [5]. The data describe one group and that group only. Much simple action research involves descriptive analysis and provides valuable information about the nature of the particular group of individuals [6]. The descriptive analysis of data provide the following:

- The first estimates and summaries, arranged in tables and graphs, to meet the objectives.
- Information about the variability or uncertainty in the data.
- Indications of unexpected patterns and observations that need to be considered when doing formal analysis.

16.1.2 Data Processing

Data Processing is simply the conversion of raw data to meaningful information through a process. Data has manipulated to produce results that lead to a resolution of a problem or improvement of an existing situation [7–20]. Similar to a production process, it follows a cycle where inputs (raw data) are fed to a process (computer systems, software, etc.) to produce output (information and insights).

16.1.3 Stages of the Data Processing Cycle

(1) **Collection** is the first stage of the cycle, and is very crucial, since the quality of data collected will impact heavily on the output. The collection process needs to ensure that the data gathered are both defined and accurate, so that subsequent decisions based on the findings are valid. This stage provides both the baseline from which to measure, and a target on what to improve.

    Some types of data collection include census (data collection about everything in a group or statistical population), sample survey (collection method that includes only part of the total population), and administrative by-product (data collection is a byproduct of an organization's day-to-day operations).

(2) **Preparation** is the manipulation of data into a form suitable for further analysis and processing. Raw data cannot be processed and must be checked for accuracy. Preparation is about constructing a dataset from one or more data sources to be used for further exploration and processing. Analyzing data that has not been carefully screened for problems can produce highly misleading results that are heavily dependent on the quality of data prepared.

(3) **Input** is the task where verified data is coded or converted into machine readable form so that it can be processed through a computer. Data entry is done through the use of a keyboard, digitizer, scanner, or data entry from an existing source. This time-consuming process requires speed and accuracy. Most data need to follow a formal and strict syntax since a great deal of processing power is required to breakdown the complex data at this stage. Due to the costs, many businesses are resorting to outsource this stage.

(4) **Processing** is when the data is subjected to various means and methods of manipulation, the point where a computer program is being executed, and it contains the program code and its current activity. The process may be made up of multiple threads of execution that simultaneously execute instructions, depending on the operating system. While a computer program is a passive collection of instructions, a process is the actual execution of those instructions. Many software programs are available for processing large volumes of data within very short periods.

(5) **Output and interpretation** is the stage where processed information is now transmitted to the user. Output is presented to users in various report formats

like printed report, audio, video, or on monitor. Output need to be interpreted so that it can provide meaningful information that will guide future decisions of the company.

(6) **Storage** is the last stage in the data processing cycle, where data, instruction and information are held for future use. The importance of this cycle is that it allows quick access and retrieval of the processed information, allowing it to be passed on to the next stage directly, when needed. Every computer uses storage to hold system and application software.

16.2 Challenging Applications

It has been well-established that a substantial competitive advantage can be obtained by data mining in general and predictive modeling in particular. For some applications, maximizing accuracy or another utility measure is of paramount importance, even at the expense of weaker explanatory capabilities. This paper examine few of the challenging application areas: insurance, fraud detection and text categorization.

(i) **Insurance**

Risk assessment is at the core of the insurance business, where actuarial statistics have been the traditional tools to model various aspects of risk such as accident, health claims, or disaster rates, and the severity of these claims. The claim frequency is rare and probabilistic by nature. For instance, the auto accident rate of an insured driver is never a clear no-accident class versus accident class problem. Instead it is modeled as a Poison distribution. The claim amounts usually follow a log-normal distribution which captures the phenomenon of rare but very high damage amounts.

(ii) **Text Categorization**

Electronic documents or text fields in databases are a large percentage of the data stored in centralized data warehouses. Text mining is the search for valuable patterns in stored text. When stored documents have correct labels, such as the topics of the documents, then that form of text mining is called text categorization. In many text storage and retrieval systems, documents are classified with one or more codes chosen from a classification system. For example, news services like Reuters carefully assign topics to news-stories. Similarly, a bank may route incoming e-mail to one of dozens of potential response sites. Originally, human-engineered knowledge-based systems were developed to assign topics to newswires. Such an approach to classification may have seemed reasonable, but the cost of the manual analysis needed to build a set of rules is no longer reasonable, given the overwhelming increase in the number of digital documents. Instead, automatic procedures are a realistic alternative, and researchers have proposed a plethora of techniques to solve this problem. The use of a standardized collection of documents for analysis and testing, such as the Reuters collection of newswires for the year 1987, has allowed researchers to measure progress in this field. Substantial improvements in automated performance have been made since then. Many automated prediction methods exist for extracting patterns from sample cases. In text mining, specifically text categorization, the raw cases are individual documents. The documents are encoded in terms of features in some numerical form, requiring a transformation from text to numbers. For each case, a uniform.

(iii) **Fraud Detection**

Fraud detection is an important problem because fraudulent insurance claims and credit card transactions alone cost tens of billions of dollars a year. In the case of credit card fraud, artificial neural-networks have been widely-used by many banks. Frauds are relatively rare, i.e. a skewed distribution that bays many traditional data mining algorithms unless stratified samples are used in the training set. Some large banks add to the transaction data volume by millions of transactions per day. The cost of processing a fraud case, once detected, is a significant factor against false positive errors while undetected fraud adds the transaction cost in the loss column. This not only influences the decision whether to declare a transaction to be processed as a fraud or not, but also calls for a more realistic performance measure than traditional accuracy. The pattern of fraudulent transactions varies with time, requiring relatively frequent and rapid generation of new models. The JAM System (Java Agents for Meta-Learning) is a recent approach for credit card fraud detection. The massive set of data with binary labels of fraud or legitimate transactions is divided into smaller subsets, for each participating bank unit and for multiple samples to gain better performance.

16.3 Predictive Model and Online Fraudulent

The typical data mining process during model creation involves the following steps. This modelling can be further explained with four distinct phases as outlined below.

1. Collecting data into a dataset that will be used for analysis.
2. Creating, training and validating a predictive model against the dataset and its known outcomes.
3. Applying the model to a new dataset with unknown outcomes.

There is a large number of credit card payments take place that is targeted by fraudulent activities shown in Fig. 16.1. Companies which are responsible for the processing of electronic transactions need to efficiently detect the fraudulent activity to maintain customers' trust and the continuity of their own business. In this paper, the developed algorithm detects credit card fraud. Prediction of any algorithm is based on certain attribute like customer's buying behaviour, a network of merchants that customer usually deals with, the location of the transaction, amount of transaction, etc. But these attribute changes over time. So, the algorithmic model needs to be updated periodically to reduce this kind of errors. Proposed System provides two

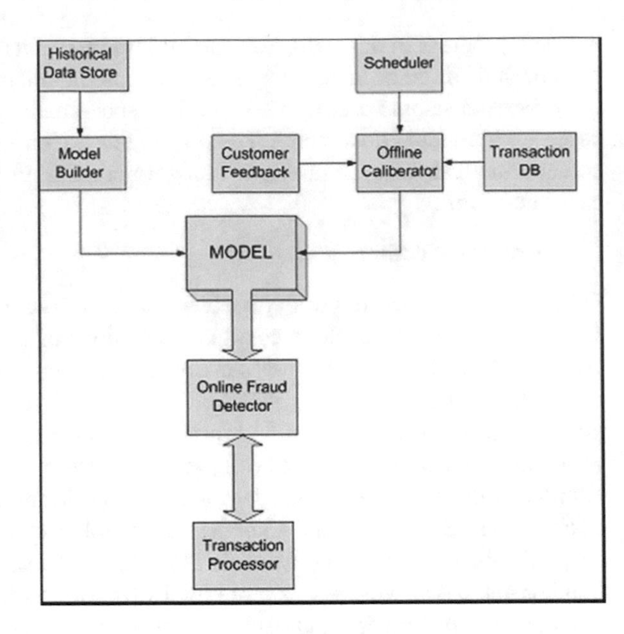

Fig. 16.1 Predictive model and online fraudulent

solutions for handling concept drift. One is an Active solution and another one is Passive. Active solution refers to triggering mechanisms by explicitly detecting a change in statistics. Passive solution suggests updating the model continuously in order to consider newly added records.

16.4 Descriptive Analysis

In the beginning, we used to primarily build models based on Regression and Decision Trees. These are the algorithms which are mostly focusing on interest variable and finding the relationship between the variables or attributes. Introduction of advanced machine learning tools made this process easy and quicker even in very complex computations.

Data Treatment: This is the most important step in generating an appropriate model input, So, we should have a smart way to make sure it's done correctly. Here are two simple tricks which you can implement:
Create dummy flags for missing value(s): In general, once we discover the missing values in a variable, that can also sometimes carry a good amount of information. So, we can create a dummy flag attribute and use those in the model.
Impute missing value with mean/any other simple value: In basic scenarios, the imputation of 'mean' or the 'median' works fine for the first iteration in a specific situation. In other cases, where there is a complex data with trend, seasonality and lows/highs, you probably need a more intelligent method to resolve for missing values.

Data Modelling: Generalized Boosting Modules (GBM) can be extremely effective for 100,000 observation cases. In cases of larger data, you can consider running a Random Forest. The below cheat-sheet will help you to decide which method to use and when.

16.4.1 Estimation of Performance of the Model

The problem of predictive modelling is to create models that are good at making predictions on new unseen data.

Therefore, it is critically important to use robust techniques to train and evaluate your models on your available training data. The more reliable your performance estimation, the more accurate the model.

There are many model evaluation techniques that you can try in **R-Programming or Python**. Below are some of them:

Training Dataset: Prepare your model on the entire training dataset, then evaluate the model on the same dataset. This is generally problematic because a perfect algorithm could skew this evaluation technique by simply memorizing (storing) all training patterns and achieve a perfect score, which would be misleading.
Supplied Test Set: Split your dataset manually using another program. Prepare your model on the entire training dataset and use the separate test set to evaluate the performance of the model. This is a good approach if you have a large dataset (many tens of thousands of instances).
Percentage Split: Randomly split your dataset into a training and a testing partitions each time you evaluate a model. It's is usually in the split ratio of 70–30% of the data. And this can give you the more significant estimate of performance and like using a supplied test set is preferable only when you have a large dataset.
Cross-Validation: Split the dataset into k-partitions or folds. Train a model in all possible aspects of data except one that is held out as the test set, then repeat this process creating k-different models and give each fold a chance of being held out as the test set. Then calculate the average performance of all k models.

This is one of the traditional and standard methods for evaluating model performance, but yeah somewhat time-consuming and has to create n-number of models to achieve the accuracy.

16.5 Experimental Results

The performance of the proposed approach is evaluated by using the credit card transaction data set using the software MATLAB. The performance of the proposed algorithm is based on the following factors

(1) Precision (2) Recall (3) F-Measure.

16.5.1 Precision Comparison

This graph shows the precision rate of existing and proposed system based on two parameters of precision and the number of Dataset. From the graph we can see that, when the number of number of Dataset is advanced the precision also developed in proposed system but when the number of number of Dataset is improved the precision is reduced somewhat in existing system than the proposed system. From this graph we can say that the precision of proposed system is increased which will be the best one. The values are given in Table 16.1.

In this graph we have chosen two parameters called number of Dataset and precision which is help to analyse the existing system and proposed systems. The precision parameter will be the Y axis and the number of dataset parameter will be the X axis. The blue line represents the existing system and the red line represents the proposed system. From this graph we see the precision of the proposed system is higher than the existing system. Through this we can conclude that the proposed system has the effective precision rate. It is demonstrated in Fig. 16.2.

Table 16.1 Number of dataset versus precision

S. No.	Number of dataset	MSHMM	SHMM
1.	10	0.3	0.21
2.	20	0.52	0.45
3.	30	0.63	0.56
4.	40	0.69	0.6
5.	50	0.75	0.64
6.	60	0.8	0.71

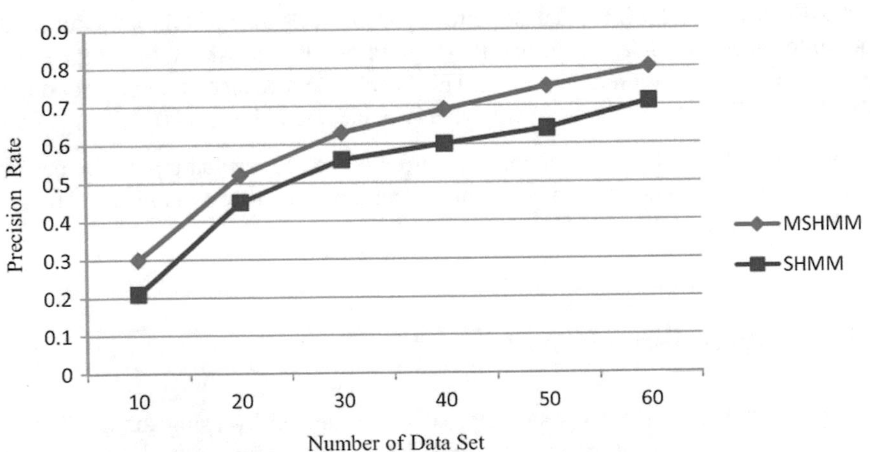

Fig. 16.2 Precision

16.5.2 Recall Comparison

This graph shows the recall rate of existing and proposed system based on two parameters of recall and number of Dataset. From the graph we can see that, when the number of number of Dataset is improved the recall rate also improved in proposed system but when the number of number of Dataset is improved the recall rate is reduced in existing system than the proposed system. From this graph we can say that the recall rate of proposed system is increased which will be the best one. The values of this recall rate are given in Table 16.2 and Fig. 16.3

In this graph we have chosen two parameters called number of Dataset and recall which is help to analyze the existing system and proposed systems on the basis of recall. In X axis the Number of dataset parameter has been taken and in Y axis recall parameter has been taken. From this graph we see the recall rate of the proposed system is in peak than the existing system. Through this we can conclude that the proposed system has the effective recall.

Table 16.2 Number of dataset versus recall

S. No.	Number of dataset	MSHMM	SHMM
1.	10	0.35	0.23
2.	20	0.43	0.32
3.	30	0.5	0.42
4.	40	0.59	0.49
5.	50	0.72	0.6
6.	60	0.87	0.69

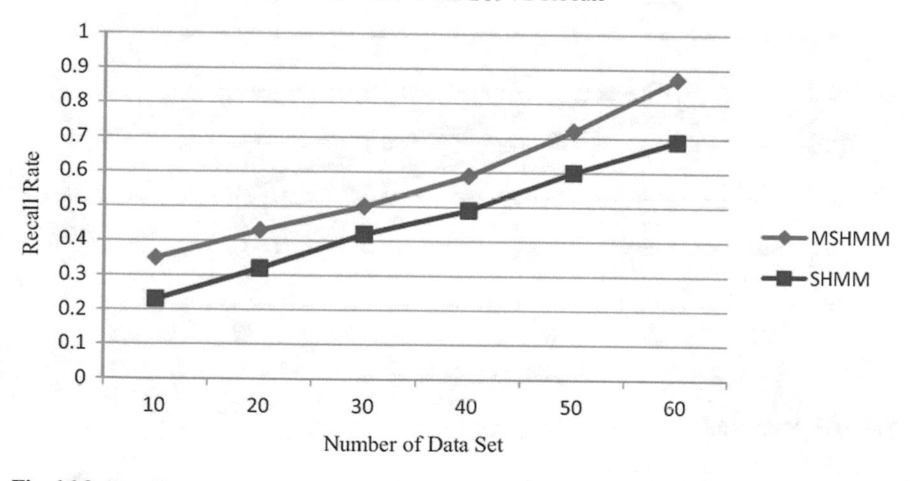

Fig. 16.3 Recall

16.5.3 F-Measure Comparison

This graph shows the F measure rate of existing and proposed system based on two parameters of F measure and number of Dataset. From the graph we can see that, when the number of number of Dataset is improved the F measure rate also improved in proposed system but when the number of number of Dataset is improved the F measure rate is reduced in existing system than the proposed system. From this graph we can say that the F measure rate of proposed system is increased which will be the best one. The values of this F measure rate are given in Table 16.3 and Fig. 16.4.

In this graph we have chosen two parameters called number of Dataset and recall which is help to analyse the existing system and proposed systems on the basis of F measure. In X axis the Number of dataset parameter has been taken and in Y axis F measure parameter has been taken. From this graph we see the F measure of the proposed system is in peak than the existing system. Through this we can conclude that the proposed system has the effective F measure.

Table 16.3 Number of dataset versus F-measure

S. No.	Number of dataset	MSHMM	SHMM
1.	10	0.89	0.76
2.	20	0.82	0.71
3.	30	0.74	0.62
4.	40	0.65	0.52
5.	50	0.54	0.42
6.	60	0.42	0.32

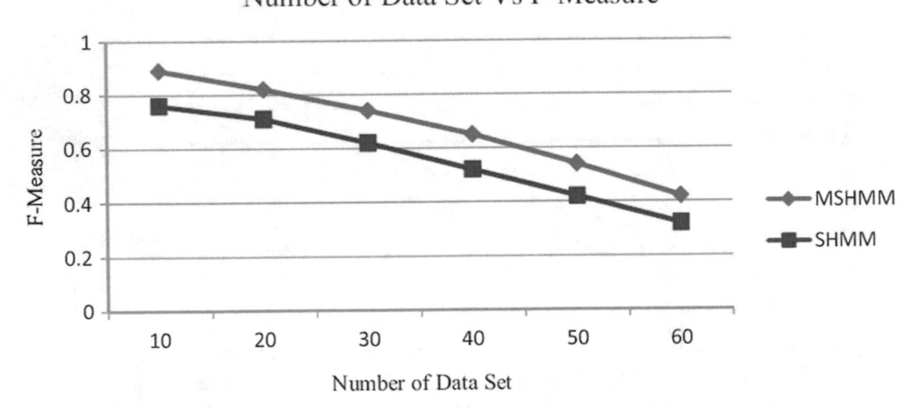

Fig. 16.4 F-measure

16.6 Conclusion

Ultimately, the work that goes into selecting algorithms to help to predict future trends and events is worthwhile. It can result in better customer service, improved sales, and better business practices. Each of these things can, of course, result in increased profits or lowered expenses. Both are desirable outcomes. The information above should act as a bit of a primer on the subject for those new to using analytics.

References

1. Brameier, M., Banzhaf, W.: A Comparison of Linear Genetic Programming and Neural Networks in Medical Data Mining. Fachbereich Informatik University at Dortmund 44221 Dortmund, Germany
2. Nguyen, T.T., Davis, D.N.: Predicting Cardio Vascular Risk Using Neural Net Techniques. University of Hull, Hull
3. Harold Robinson, Y., Golden Julie, E.: MTPKM: Multipart trust based public key management technique to reduce security vulnerability in mobile ad-hoc networks. Wirel. Pers. Commun. **109**, 739–760 (2019)
4. Harold Robinson, Y., Santhana Krishnan, R., Golden Julie, E., Kumar, R., Son, L.H., Thong, P.H.: Neighbor knowledge-based rebroadcast algorithm for minimizing the routing overhead in mobile ad-hoc networks. Ad Hoc Netw. **93**, 1–13 (2019)
5. Gao, D.W., Wang, P., Liang, H.: Optimization of hidden nodes and training times in ANN-QSAR model. College of Forest Resources and Environment, Northeast Forestry University, Harbin 150040, China, School of Municipal and Environmental Engineering, Harbin Institute of Technology, Harbin 150090, Chin
6. Davis, D.N., Nguyen, T.T.T.: Generating and verifying risk prediction models using data mining (A case study from cardiovascular medicine). Department of Computer Science, University of Hull, Cottingham Road, Hull, HU6 7RX, UK
7. Carneiro, E.M.: Cluster analysis and artificial neural networks: a case study in credit card fraud detection. In: 2015 IEEE International
8. Al-Jumeily, D.: Methods and Techniques to Support the Development of Fraud Detection System. IEEE (2015)
9. Mishra, M.K.: A comparative study of Chebyshev functional link artificial neural network, multi-layer perceptron and decision tree for credit card fraud detection. In: 2014 13th International Conference on Information Technology (2014)
10. Balaji, S., Golden Julie, E., Harold Robinson, Y.: Development of fuzzy based energy efficient cluster routing protocol to increase the lifetime of wireless sensor networks. Mobile Netw. Appl. **24**(2), 394–406 (2019)
11. Mareeswari, V.: Prevention of credit card fraud detection based on HSVM. In: 2016 IEEE International Conference on Information Communication and Embedded System (2016)
12. Assis, C.A.S.: A genetic programming approach for fraud detection in electronic transactions. In: 2015 Second International Conference on Advances in Computing and Communication Engineering (ICACCE) (2015)
13. Harvey, D.Y.: Automated feature design for numeric sequence classification by genetic programming. IEEE Trans. Evol. Comput. **19**(4) (2015)
14. Balaji, S., Golden Julie, E., Harold Robinson, Y., Kumar, R., Thong, P.H., Son, L.H.: Design of a security-aware routing scheme in mobile ad-hoc network using repeated game model. Comput. Stand. Inter. **66**, (2019)
15. Sahin, Y., Bulkan, S., Duman, E.: A cost-sensitive decision tree approach for fraud detection. Expert Syst. Appl. **40**, 5916–5923 (2013)

16. Harold Robinson, Y., Balaji, S., Golden Julie, E.: FPSOEE: Fuzzy-enabled particle swarm optimization-based energy-efficient algorithm in mobile ad-hoc networks. J. Intell. Fuzzy Syst. **36**(4), 3541–3553 (2019)
17. Van Vlasselaer, V.: APATE: a novel approach for automated credit card transaction fraud detection using network-based extensions. Published in Decision Support Systems (2015)
18. Kültür, Y.: A novel cardholder behavior model for detecting credit card fraud. In: IEEE International Conference on Commuting and Communication Engineering (2015)
19. Harold Robinson, Y., Rajaram, M.: Energy-aware multipath routing scheme based on particle swarm optimization in mobile ad hoc networks. Sci. World J., 1–9 (2015)
20. Harold Robinson, Y., Rajaram, M.: A memory aided broadcast mechanism with fuzzy classification on a device-to-device mobile ad hoc network. **90**(2), 769–791 (2016)

Chapter 17
Design and Analysis of Wearable Patient Monitoring System (WPMS)

Pankaj Singh, Anand Saunil Shingwekar, Mansi Sharma and Shubham Vyas

Abstract Wearable patient monitoring system (WPMS) used for examining the health of the patient, was first introduced in the year 1950. When the Holter monitoring system was introduced as one of the clinical utilisation, but it does not prove out to be successful, efficient and reliable due to lack of techs and innovations. Over the last few decades, lot of advancement is taken in the field of sensors and systems. In this paper, we will be analysing specific body parameters which are Heart rate (HR), Blood pressure (BP), temperature. These parameters are examined with the use of body sensors which are attached to the body. These sensors are interfaced with a μC with other components to help in the proper examination of the parameters of our body. This product is capable of communicating to the end-user (qualified person) with the help of ZigBee Wi-Fi, GSM shield by creating a Wireless sensor network (WSN) (Custodio et al. in Sensors 12:13907–13946, 2012 [1]) and transmitting data from the sensor to the μC using Body area network (BAN) (Custodio et al. in Sensors 12:13907–13946, 2012 [1]).

Keywords Wearable patient monitoring system (WPMS) · Heart rate (HR) · Blood pressure (BP) · μC (Microcontroller) · Wireless sensor network (WSN) · Body area network (BAN)

17.1 Introduction

With the advancement of new technologies, such as AI, Machine learning and big data, medical wearable has added a great value to the health care with the special emphasis on early diagnosis, treatment, patient monitoring and prevention of risky cardiovascular disease (CVD) and chronic diseases. Medical wearable technology equipped with the tele-medicines system (TMS) is a new emerging technology in which, the essentials body parameters of the patient which include both physiological and non-physiological activity [1] could be assessed with the help of body sensor networks (BSN) which includes wired/wireless wearable sensors.

P. Singh (✉) · A. S. Shingwekar · M. Sharma · S. Vyas
SRM Institute of Science and Technology, Modinagar 201204, India
e-mail: singhpankaj76@gmail.com

V. E. Balas et al. (eds.), *Internet of Things and Big Data Applications*, Intelligent Systems Reference Library 180, https://doi.org/10.1007/978-3-030-39119-5_17

These sensors would be attached to the patient's body to measure essential body parameters. With the help of WSN [1], the sensed data could be transmitted to the end-user or the qualified person. BAN and WSN [1] are two important components of our WPMS. This WPMS or our product proves out to be efficacious and extremely useful for the aged patients and people living in remote areas or locations lacking EMERGENCY medicals facilities. The future of wearable tech is very bright and challenging [2]. The use of a wearable medical device is expected to reach about 12.1 billion by 2023.

The product is vital in case of an EMERGENCY. The information is sent to the doctor if the device is in offline mode through SMS. The device is connected to the internet with a hotspot created by a smartphone near the patient [3].

BSN and WSN together work as a single unit which assists the device to sense the physical signal of patient's body and converting it into an electrical signal that is further transferred to the end-user in an application or a website, where the doctors can monitor the condition of the patient [4].

This paper deals with the working of the WPMS and its test results.

17.2 Literature Review

During the last few decades, a lot of advancement has been seen in the field of smart wearable technology. This technology allows the individual to keep a track of his personal health status and in case of critical/abnormality, the information could be sent to the hospital or an authorized and qualified person for an examination of the situation. It is also cost and time effective as a person does not have to pay a visit to the hospital multiple times for a checkup. In the wearable medical system, a group of wired and wireless sensors are being used to assess the health of a patient or a person.

The functionality of sensors is to collect the data in the physical form and transmit the data after converting into electrical signals to the desired destination.

Basically, the health monitoring system that is available in very bulky in size and therefore a patient has to visit the hospital time and time again for regular check-ups. To eliminate this problem patient monitoring system is designed so that the person could assess his health timely. The two modern-day technology includes the use of wireless and the wearable technology which would assess the health of the person and according maintain his health and fitness. There are certain disadvantages associated with these devices as they are not accurate and they are bulky in size. A lot of research work is undergoing in order to remove the ambiguity which is associated with these devices.

A number of wearable systems to assess the health of the patient is available in the market. The companies which are working in developing the health-care devices are Fitbit and Garmin so that it provides much more accessible to the patient in using these health-care devices [4]. These devices are built in such a manner that it would able the analyze the vital body parameters as efficiently as the devices which are present in the hospital.

17.3 Theoretical Framework

The four essentials that specify the health status of an individual and should be monitored regularly are body temperature, respiration rate, blood pressure, pulse rate, body movement. These four parameters of the human body vary according to age, sex, weight and height. The spectrum of a healthy person in terms of the above parameters mentioned is being illustrated in Table 17.1.

The death rate in the world due to cardiac vascular diseases is very significant which is about 17.9 million per year that account for approximately 31% of the total deaths [5]. In every 43 s, a person dies due to heart disease. The main reason for deaths due to heart diseases would be excessive intake of alcohol, taking too much of stress, intake of poor dietary food, diabetes, smoking, high blood pressure.

The difference between a heart attack and cardiovascular disease is when the blood flow from coronary arteries are blocked is know to have a heart attack whereas when the heart stops pumping blood around the body then it is known as cardiac arrest.

Wearable technology proves out to be useful in preventing heart diseases. According to recent studies, about 75% of the cardiovascular diseases could be prevented by continuous tracking of the heart rate and to take essential steps and decisions to prevent the heart diseases and save a life [6].

Table 17.1 Normal range of body parameters

Parameters	Normal range (Men)	Normal range (Woman)
Body temperature	36.7 °C	36.2°C
Respiration rate	12–18 Breadths per minutes	12–18 Breadths per minutes
Heart rate	At rest 70–80 Beats per minutes	At rest 70–80 Beats per minutes
Blood pressure	Systolic BP	110–130 mm Hg
Blood pressure	Diastolic BP	70–90 mm Hg

17.4 Product Requirements and Design

17.4.1 Product Requirement

This section deals with the requirements of the product and requirements in the specific order to be implemented.

- To interface required sensors with E-Healthkit
  - ArduinoMega interface
    E-HealthKit and its Libraries
    ArduinoMega μC & its IDE
- To enable required body parameters to be transferred through text messages or internet
  - Networking tools
    Wifi-Module
    Wifi-communications
    GSM-shield
  - Phone
    An Android Smart phone
- Connection b/w Android phone and Arduino MiddleWare
  - Portals
    Zig-bee, Wifi-module, supports IEEE 802.15.4
    Android smart phone's hotspot
- A remote server, storing essential body parameters
  - Database systems
    MYSQL data base
- To enable qualified person to obtain body parameters of specific patient
  - A website or an Application
    Dashboard to obtain data using MYSQL database for details.

17.4.2 Design

In our product, there are three essential phases which include firstly, the body sensors which measure thc essential parameters of the body i.e temperature, heart rate and blood pressure. Further the signal is being converted into electrical signal and the information is send to the central component i.e Arduino μC. The health status of an individual is been transferred to the qualified person through SMS. In the next

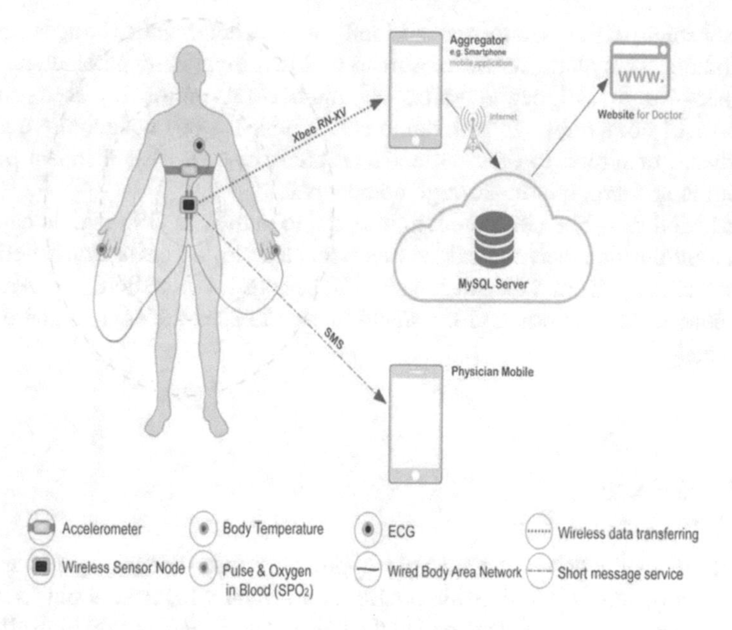

Fig. 17.1 A prototype of our product with additional sensors [7]

phase the smartphone allows data to be send to a database. Aqualified person in this, a doctor could examine his patient (Fig. 17.1).

17.5 Product Working

A product consists of E-Health wearable smart system (EHWSS) with other essential components, to obtain a useful product i.e gaining recognition and interest in the field of medical for commercial use as well as for exploring in the academic sector.

There are methods that are required to be followed for its functionality. This section presents the procedure of our product's functionality.

17.5.1 Hardware

1. ArduinoMega 2650—μC board used for transmission and reception of data by various communication techs like Wifi, bluetooth, Zigbee, IEEE 802.15.4. The Arduino board is used because ArduinoMega 2560 is capable to read input from one end-user and transmit output to other end-user.
2. E-Health Platform—Enables ArduinoMega 2560 to conduct application related to medical [8]. This platform is capable of interfacing different sensors through

sensor shield of the spectrum from blood pressure, heart rate, to body temperature andothers. This platform use sensors to pull in critical body parameters.

3. Zigbee—In the Zigbee module, communication protocol, based on IEEE 802.15.14 compatible with Arduino smartphones. Does not require additional hardware or circuit to create a personal area network. Establishes a path b/w Arduino and smartphone through hotspot path.
4. GSM Modem—The GSm modem in Arduino known as GSM shield capable of transmitting and receiving text messages that enable doctors and qualified person to receive data from a distant location. In case of EMERGENCY, if Arduino is not connected to internet, GSM shield is used to alert the doctors and qualified persons.

17.5.2 Software

1. Android application—The android application, user-friendly platform which allows a patient to monitor his health on a screen which turns out to be quiet effective as he does not have to visit the hospital and he/she could himself/herself assess his health condition. The features of the android application include the following tabs.
   (a) Chart that illustrates health status of an individual graphically.
   (b) Data which includes numerical values obtained through sensors.
   (c) About tabs specify and assess the health of the patient.
2. Doctor's portal—The portal includes the two main pages which are sign in page and the dashboard page. The sign in page consists of the user name and the password through which only the doctor could access the webpage. The dashboard page consists of all the measuring parameters of the body with the help of graphical and numerical analysis which also compares the actual body parameter, value of the person with normal body parameters and further the doctor could also prescribe the patient.

17.6 Analysis and Result

The above section aimed at evaluating the accuracy of the system to monitor the health of the patient. The system is mainly divided into three sections in which the sensor will measure the essential body parameters and the data transfer will be done using various communication technologies and doctor obtaining the data from its portal, all are independent.

17.6.1 Transmission

In this particular section, we will be assessing, how effectively a sensor and a zigbee module used for data transmission and check whether the doctor is able to obtain a medical report of his patient from an online portal. In Table 17.2, the components have been thoroughly checked according to their functionality.

Table 17.2 Sensor's functionality [7]

S. No.	Component and sensors	USE	Description
1	Body temperature sensor	Monitoring body temperature	When temp sensor is examined on its capabilities and functionality and transmission of sensed data
2	Heart rate sensor	Monitoring heart rate of patient	When heart rate sensor is examined on its capabilities and functionality and transmission of sensed data
3	BP sensor	Monitoring blood pressure of patient	When BP sensor is examined on its capabilities and functionality and transmission of sensed data
4	Zigbee module	Testing zigbee	When zigbee module is tested for its connection with hotspot on smart phone using UDP and HTTP protocol to send data in app and server
5	GSM shield	Testing GSM module	In Gsm, modem is tested in case of offline mode to send the vital parameters of patient's body to doctor
6	Website	Testing website functionality	When the website is tested in case of an EMERGENCY as well as checking the delivery of data to the doctor

Table 17.3 Evaluation of results

S. No.	Attempts	Accuracy (%)	Remark
1	10	98	PASS
2	20	60	PASS
3	20	60	PASS
4	40	50	PASS
5	5	80	PASS
6	15	93.3	PASS

17.6.2 Results and Discussions

In Table 17.3, all the different components used worked independently of each other. Each component has been tested and if the accuracy comes out to be greater than 50% then product or component is effective.

The outcome of the result illustrates that there is a vast scope of improvement that could be done in the system for achieving better results for real-time monitoring.

17.7 Conclusion

The above research work has been done to assess and develop the wearable medical instruments and devices which allows the patient to measure their specific essential body parameters and also offers the monitoring of the patient by the doctor.

With the advancement in the field of tele-medicines wearable medical devices is a platform where the doctor could examine the health status of the patient from the remote area.

At the end, the lots of research is going on to enhance and expand this system and product for real time applications effectively.

References

1. Custodio, V., Herrera, F., Lopez, G., Morena, J.: A review on architectures and communications technologies for wearable health-monitoring systems. Sensors **12**(12), 13907–13946 (2012)
2. Makikawa, M., Shiozawa, N., Okhada, S.: Wearable Sensors. Elsevier B.V., The Netherlands (2014)
3. Puddu, P.E., Ambroisi, A.D.: Systems Design for Remote Healthcare, p. 1. Springer, New York (2014)
4. Digital Health Technology Vision 20 17, 2017. Available: www.accenture.com/us-en/insight-digital-health-tech-vision-2017
5. Amft, O., Van Laerhoven, K.: What will we wear after smartphones? IEEE Pervasive Comput. **16**(4), 80–85 (2017)

6. Omoogun, M., Ramsurrun, V., Guness, S., Seeam, P., Bellekens, X., Seeam, A.: Critical patient eHealth monitoring system using wearable sensors. In: 2017 1st International Conference on Next Generation Computing Applications (NextComp), pp. 169–174 (2017)
7. Vincent, J.-L., Einav, S., Pearse, R., Jaber, S., Kranke, P., Overdyk, F.J., Whitaker, D.K., Gordo, F., Dahan, A., Hoeft, A.: Improving detection of patient deterioration in the general hospital ward environment. Eur. J. Anaesthesiol. **35**(5), 325 (2018)
8. Wu, T., Wu, F., Redoute, J.-M., Yuce, M.R.: An autonomous wireless' body area network implementation towards IoT connected healthcare applications. IEEE Access. **5**, 11413–11422 (2017)

Chapter 18
High Gain Microstrip Patch Antenna for Wireless Application

Anurag Singh, Jay Prakash Narayan Verma and Nishant Srivastava

Abstract In the paper, a circular slot high gain rectangular patch antenna is presented. A circular slot integrated patch structure is introduced to enhance the gain of proposed antenna. The antenna is designed on 6.7 mm thick RT Duroid substrate to intensify the gain. The proposed antenna has perfect broadside radiation pattern and high gain of 6.9 dBi below −10 db. The antenna is simulated using HFSS software at operating frequency of 2.4 GHz and exhibits good result.

Keywords Patch antenna · High gain · Radiation pattern · VSWR

18.1 Introduction

Microstrip patch antennas is an appealing in the theory of antenna and designing attracting many these days for its space borne applications, radars, WLAN, mobile applications, etc. Apart from the desired merits like very light weight, low profile, ease in integration with feed networks and less cost of fabrication it can also conform to any curved surface [1–3]. Low gain that is less than or equal to 2 dB and narrow bandwidth specifically around 5% in accordance to centre frequency remain to be one of the main disadvantages [2].

In order to enhance the existing technology used in communication devices there is a requirement to analyse how changes in dielectric constant of a microstrip patch antenna affects its bandwidth and the overall size [3]. There are many ways for increasing the bandwidth of antenna which includes -different impedance matching, techniques of feeding and usage of slotting antenna geometry. However, due to mutual conflicts in the properties of size and bandwidth of antenna enhancing one

A. Singh (✉) · J. P. N. Verma · N. Srivastava
Department of ECE, SRM Institute of Science and Technology, Chennai, India
e-mail: agill56@gmail.com

J. P. N. Verma
e-mail: jpnverma2@gmail.com

N. Srivastava
e-mail: nishant295@gmail.com

V. E. Balas et al. (eds.), *Internet of Things and Big Data Applications*, Intelligent Systems Reference Library 180, https://doi.org/10.1007/978-3-030-39119-5_18

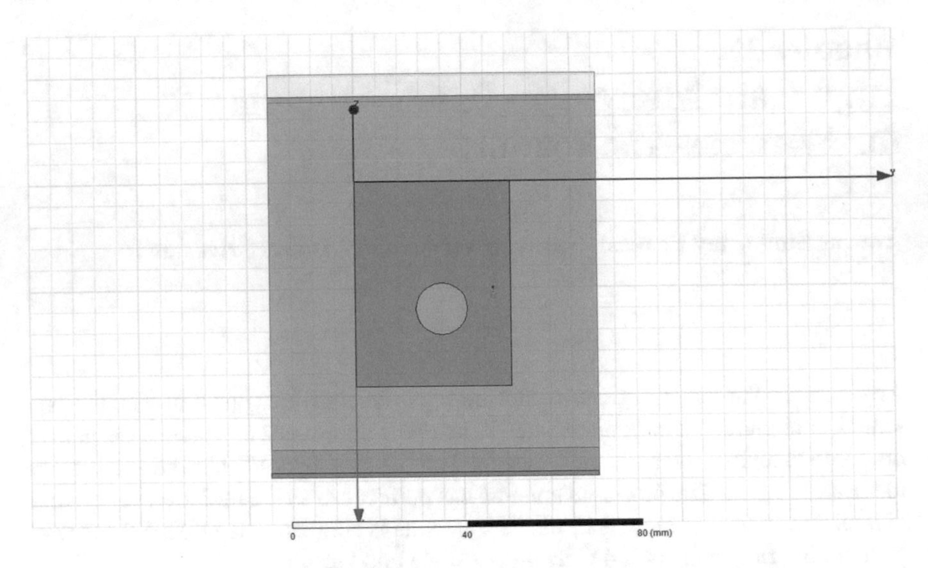

Fig. 18.1 Structure of proposed antenna

characteristics result leads to lowering of another [4, 5]. In microstrip patch antenna, the primary source of the EM radiations generated is fringing field [4–6]. The quantity of the fringing fields causing radiation is a function of height of the substrate and dimensions of the patch. The radiation can be influenced using low permittivity thicker substrate and frequency [7–11].

In this research paper, a rectangular microstrip patch antenna is designed at resonant frequency of 2.4 GHz with a circular slotting approximately in the middle of the patch. The antenna dimensions are optimized to achieve the desired result. The microstrip fed patch antenna has conducting electrical layers which are separated by 6.7 mm dielectric substrate having dielectric constant of 2.2 and dielectric substrate RT Duroid 5880 is used. For maximum EM radiation in the desired direction, a substrate with low dielectric constant is used (Fig. 18.1).

In this research paper, we have designed a rectangular microstrip patch antenna for resonant frequency 2.4 GHz with a circular slotting approximately in the middle of the patch. The dimensions are calculated as such to get the best antenna characteristic outputs.

18.2 Design and Simulation

The structure and design of thc antenna proposed is shown in Fig. 18.2 consists of three surfaces. The lower surface is ground plane and the middle layer (substrate) is made up of RT Duroid 5880 with relative permittivity of 2.2. The antenna is fed by a coaxial probe on midpoint of one of the edges of the patch. The conductor surrounding the coaxial cable is connected to floor plane, and middle conductor is prolonged on to the patch antenna. The microstrip antenna is designed for the resonant frequency

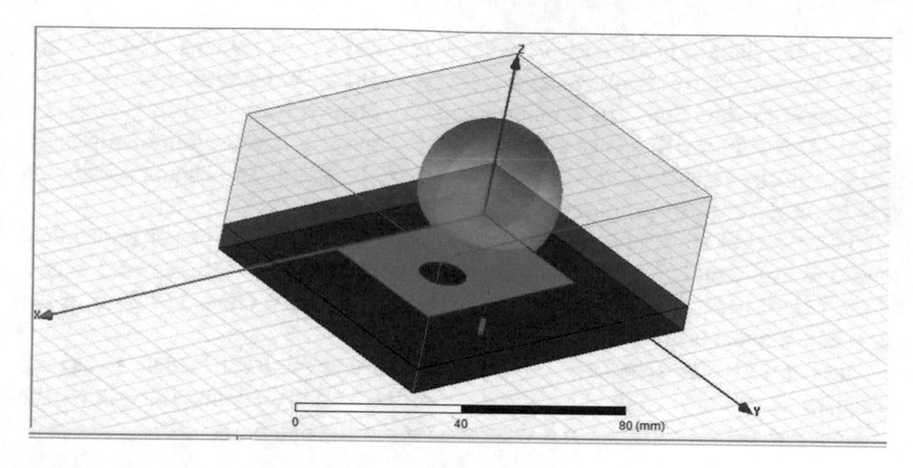

Fig. 18.2 Simulated model of proposed antenna

Table 18.1 Dimensions of patch and substrate

	Length (L) (mm)	Width (W) (mm)	Height (mm)
Patch	36	48	0
Substrate	76	88	6.7

of 2.4 GHz which is used for WIFI, LAN and several other wireless applications. The feeding technique used for feeding the antenna is coaxial probe method. The material of the patch and the ground is taken to be copper as it is easily available and cheap (Table 18.1).

Return loss is the power lost in the transmission line due to reflection or return in the signal. The obtained result of S11 parameter of the antenna is portrayed in Fig. 18.3. The return loss of designed antenna is displayed which shows that it is resonating at a frequency of 2.4 GHz.

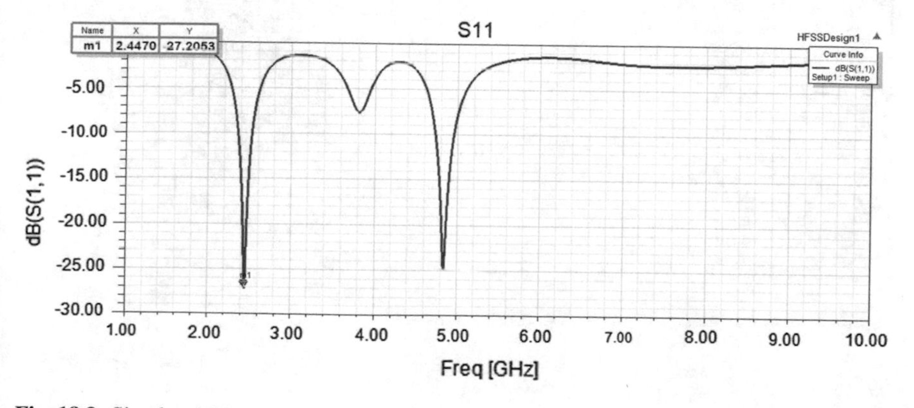

Fig. 18.3 Simulated S11 parameter (return loss)

Radiation pattern is the energy which is radiated by the antenna. The dramatic representation of the radiation pattern is the spread of the radiated energy into space, as function of direction. The radiation pattern obtained for $\varphi = 0$ to $\varphi = 360$ is analyzed. For $\Phi = 0$ and $\varphi = 90$ is of utmost important because Microstrip patch antenna mostly radiates normal to the surface of the patch. The obtained radiation pattern for the antenna proposed is shown in the Fig. 18.4.

The reception apparatus gain is characterized as how much power transmitted toward pinnacle radiation to that isotropic source. The Fig. 18.5 represents the gain of the Microstrip patch antenna at resonant frequency 2.4 GHz. The gain of the proposed antenna is recorded as 6.93 dB.

The Fig. 18.6 represent the VSWR of proposed antenna. It can be defined as ratio of maximum voltage to minimum voltage of the antenna. Figure 18.6 shows the frequency vs dB plot. The VSWR recorded for the antenna is 1.49 dB.

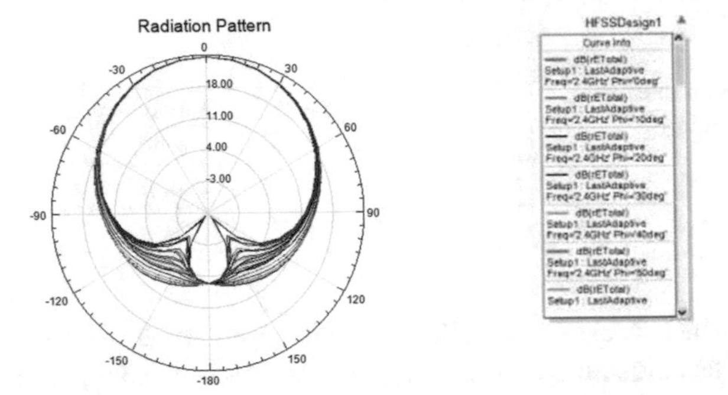

Fig. 18.4 Radiation pattern of proposed antenna

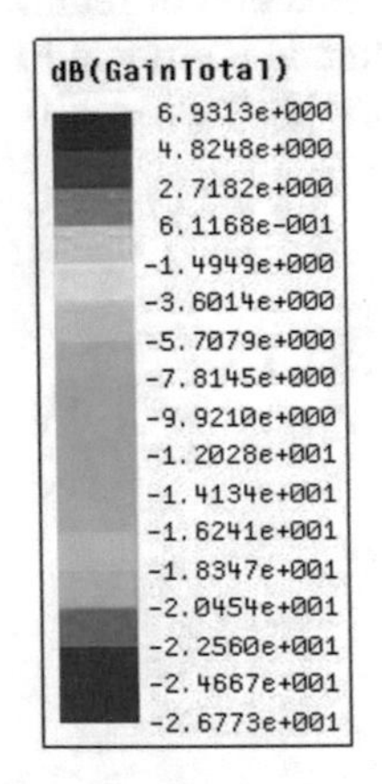

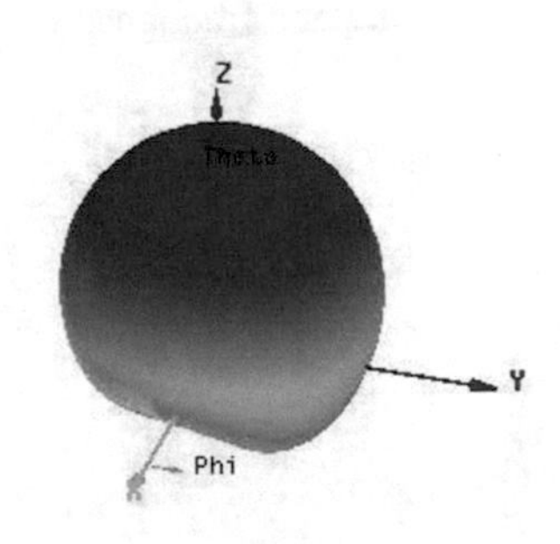

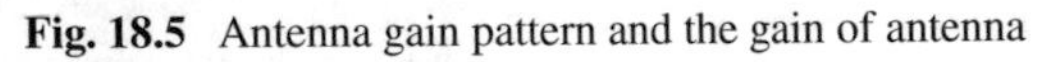

Fig. 18.5 Antenna gain pattern and the gain of antenna

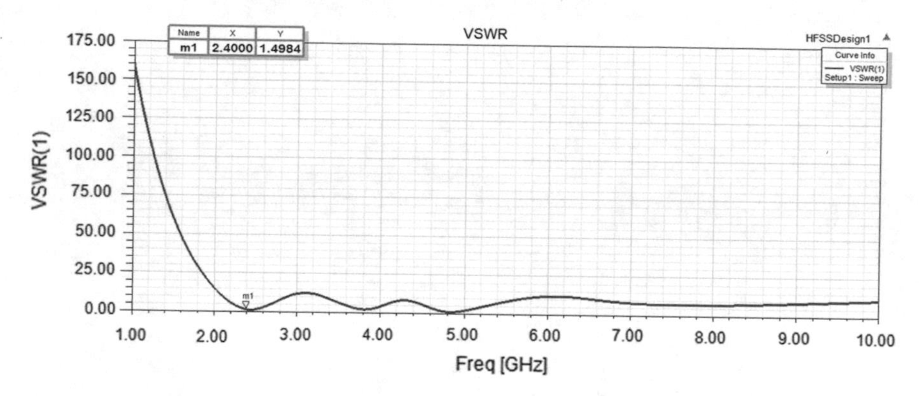

Fig. 18.6 VSWR of the proposed antenna

18.3 Conclusion

The microstrip patch antenna is designed at frequency 2.4 GHz using HFSS (High Frequency Structural Simulator). The patch and the ground are separated by RT Duroid 5880. Parameters like S11(return loss), gain, directivity, radiation pattern and VSWR was analysed, simulated and recorded. The parameters were considerably better than the existing model.

References

1. Balanis, C.A.: Antenna Theory Analysis and Design, 2nd edn. Wiley, New York (1997)
2. Pozar, DM.: Microwave Engineering, 3rd edn. Wiley, New York (2005)
3. Biradar, B.S., Sankpal, S.V.: Designing of rectangular microstrip patch antenna for wireless communication at 2.4 GHz. Int. J. Sci. Tech. Eng. **3**(8) (2017)
4. Stutzman, W.L., Thiele, G.A.: Antenna Theory and Design, 2nd edn. Wiley, Inc. (1998)
5. Shimels Mamo.: Analysis and design of dual band microstrip antennas for mobile handsets master's thesis (Unpublished 2005)
6. Kumar, P., Thakur, N.: Micro strip patch antenna for 2.4 GHZ wireless applications. Int. J. Eng. Trends Tech. **4**(8) 2013
7. Karthick, M.: Design of 2.4 GHz patch antennae for WLAN applications. In: IEEE Seventh National Conference on Computing, Communication and Information System (2015)
8. Mokha, S., Jangir, S.: Design of flexible microstrip patch antenna of 2.4 GHz operation frequency using HFSS. Int. Res. J. Eng. Tech. **3**(5), (2016)
9. Chandrakanta, K., Debatosh, G.: Asymmetric geometry of defected ground structure for rectangular microstrip: a new approach to reduce its cross—polarized fields. IEEE Trans. Antennas Propag. **64**, 2503–2506 (2016)
10. Collin, R.E.: Antennas and radio wave propagation. McGraw-Hill Inc, New York (1985)
11. Wong, K.L.: Planar antennas for wireless communications. Wiley, New York (2005)

Chapter 19
Wi-Fi Based Inertial RSS and Fingerprinting Using Multi-agent Technology

M. Subramanian, Y. Harold Robinson and A. Essakimuthu

Abstract First, the costly offline process of RSS map construction in conventional fingerprint-based localization is removed by the use of inertial sensors. To collect RSS fingerprints automatically and proposed system conducts inertial sensor-based self-localization and estimates. Second, we successfully constructed the collective fingerprint map by a credibility-based user collaboration scheme. The proposed system is based on a passive user inertial method, which collects and uploads fingerprints automatically in daily life using MAS. We show how to handle multiple floors and stairways, how to handle symmetry in the environment, and how to initialize the localization algorithm using Wi-Fi signal strength to reduce initial complexity.

Keywords Inertial sensors · Location fingerprinting · Multiagent technology (MAT) · Wi-Fi-based fingerprinting · Received signal strength (RSS)

19.1 Introduction

Locating mobile objects is an essential function needed by location-aware applications in pervasive computing environments [1, 2]. The Global positioning system (GPS) [3] has commonly been used in outdoor environments and been widely adopted in modern mobile devices such as smart phones. In indoor environments, however, no outstanding solution has been found due to practical issues which are related to complicated infrastructure requirements. Conventional mechanisms for indoor node localization are based on various types of infrastructure support, which include received signal strength (RSS) fingerprints [4], ultra wideband (UWB) [5],

M. Subramanian · A. Essakimuthu
Department of Computer Science and Engineering, SCAD College of Engineering and Technology, Cheranmahadevi, India
e-mail: m.subramanian86@gmail.com

A. Essakimuthu
e-mail: karthic.tvl@gmail.com

Y. Harold Robinson (✉)
School of Information Technology and Engineering, Vellore Institute of Technology, Vellore, India
e-mail: yhrobinphd@gmail.com

V. E. Balas et al. (eds.), *Internet of Things and Big Data Applications*, Intelligent Systems Reference Library 180, https://doi.org/10.1007/978-3-030-39119-5_19

ultrasound [6], radiofrequency identification (RFID) [7], inertial measurement units (IMUs) [8], etc. Among the diverse approaches for indoor node localization, the RSS-based fingerprinting system is considered practical since the system can easily be deployed using the current wireless (i.e., IEEE 802.11 Wi-Fi infrastructure).

For the past few years, active research has been conducted on localization methods based on the RSS fingerprints. RADAR [9] was the first Wi-Fi-based localization system which utilized RSS from APs in the vicinity. Ekahau [10] commercialized the concept with enhanced localization accuracy. These systems, however, require an RSS map-building process via laborious offline training for a specific site of interest. Place Lab [11] and Harold [12] tried to reduce the cost for offline training by collecting the Wi-Fi signals automatically while moving with a GPS-equipped vehicle. These systems are primarily targeted for outdoor environments and are certainly not applicable to indoor applications. Redpin [13] handled the offline training by incorporating a collaborative approach where users collect fingerprints while using the devices. The system mandates active participation of users to collect fingerprints because users must manually register new fingerprints frequently to enhance the location accuracy. The methodology [14] has reduced the number of fingerprints by clustering the RSS signals with similar measurement values. Also, several studies were conducted to improve the positioning accuracy of the RSS fingerprint-based mechanism. Bayesian modeling [15] and statistical learning [16] belong to this category. Meanwhile, many IMU-based localization systems have been developed, especially in the robotics area, to track the locations of mobile robots [17]. Applying the technique to human tracking poses a new challenge since human behavior, compared to robots, is dynamic. Moreover, a subtle motion of human movement should be carefully considered to obtain reasonable tracking accuracy. The navigation system [18] proposed a pedestrian navigation system that achieves accuracy of 2 m both indoors and outdoors by integrating GPS and IMU. The new localization system which uses an accelerometer for step counting and a magnetometer to track [19, 20].

Several issues should be considered for the practical use of the RSS fingerprint-based localization. In particular, constructing a high-quality RSS fingerprint map is an essential part of the system since localization accuracy highly depends on fingerprint quality. The RSS map-building process typically requires an extensive and thorough site surveying, usually done manually with specific hardware and software tools. Much effort has recently been given to reducing the cost and complexity of fingerprint map building.

19.2 System Overview

The proposed system consists of mobile users and the fingerprint server. The mobile users automatically collect the RSS fingerprints for Wi-Fi APs in the vicinity while localizing their position based on their smart phones. The fingerprint server constructs a collective RSS map by integrating individual fingerprints received from the mobile users; then it supplies the map to newly entered users for localization or even back to

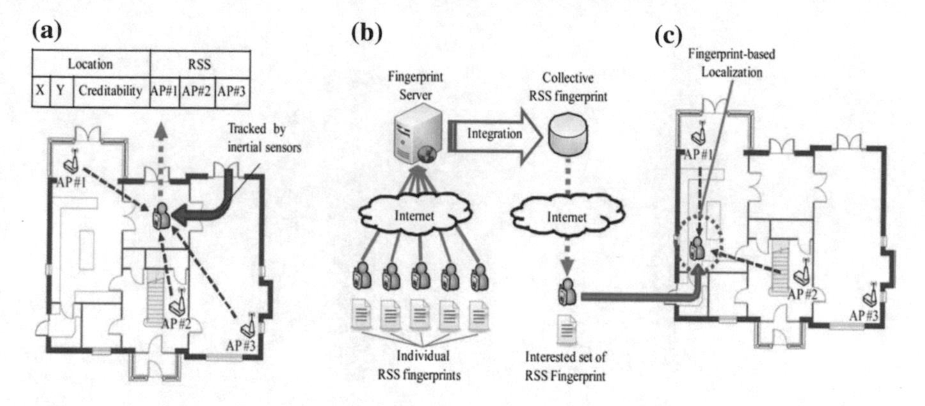

Fig. 19.1 System overview. **a** RSS fingerprint collection. **b** Collective RSS fingerprint construction. **c** Fingerprinting-based localization

the mobile users to further enhance the localization accuracy. Figure 19.1 illustrates the overall structure of the proposed system. The localization of the mobile object is achieved in three steps: local collection of the RSS fingerprints, construction of the global RSS fingerprints map, and the localization process.

We first propose an Itinerary Energy Minimum for First-source-selection (IEMF) algorithm, which extends LCF by considering the estimated communication cost.

19.3 Proposed Multi-agent Itinerary Planning Algorithms

19.3.1 *Estimated Communication Cost of a Candidate Itinerary*

We first show how to estimate the communication cost of a given itinerary $t|S[1]|S[2]|$ $\cdots$ $|S[n]|t$, which means that an agent starts from sink t and returns back to t after visiting n source nodes, as shown in Fig. 19.1. Generally, the communication energy consumption for a packet transmission at a given node consists of the receiving energy, the control energy, and the transmitting energy. Let $e$ctrl be the energy spent on control messages exchanged for a successful data transmission (e.g., acknowledgement). Let $m$rx and $m$tx be the energy consumption for receiving and transmitting a data bit, respectively (Fig. 19.2).

Let $c$tx denote the fixed energy cost for each transmission, which is independent of the packet length. Without loss of generality, we assume that $m$tx, $m$rx, $c$tx, and $e$ctrl are the same for every node. Let $l$rx and $l$tx be the sizes of a received and a transmitted packet, respectively. When a node receives a packet with size of $l$rx, after local processing, the size of a transmitted packet by this node ($l$tx) is different. The communication energy consumption at a node (i.e., an intermediate node or a source node) can be expressed as

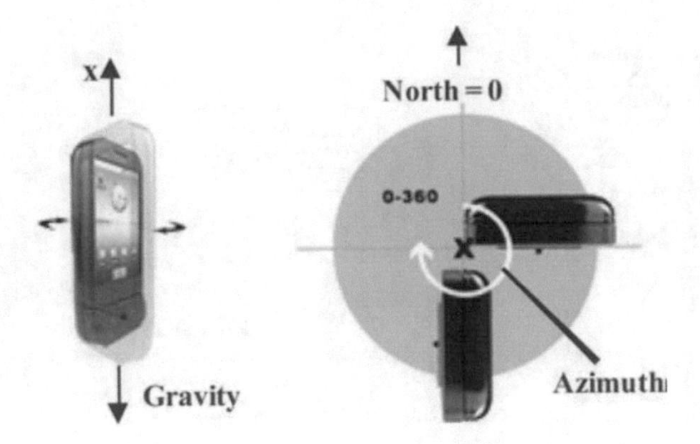

Fig. 19.2 Overall process to determine the direction of movement axis. We first find the relative horizontal plane to the ground and determine the movement axis by the initial direction of movement

$$e(l_{rx}, l_{tx}) = m_{rx}l_{rx} + (m_{tx}l_{tx} + c_{tx}) + e_{ctrl}$$

Multiple hops may exist between two adjacent source nodes, e.g., $S[k-1]$ and $S[k]$. Let $d(S[k-1], S[k])$ denote the distance between the two source nodes. In a dense WSN, we can estimate the hop count between $S[k-1]$ and $S[k]$ as $Hk\ k-1=$ _$d(S[k-1], S[k])/R$_, where R represents the maximum transmission range. When the agent traverses intermediate sensor nodes (not the source nodes), the agent size remains the same.

19.4 Single-Agent Itinerary Planning Problem

19.4.1 Data Aggregation Model

The degree of correlation between sensory data from two sensor nodes is closely related to the distance between them, as well as a particular application scenario. Typically, closely located sensors are very likely to generate data with high redundancy. In densely populated WSNs, data aggregation becomes a very important function for energy conservation, which reduces the redundancy in sensor data and, thus, decreases the volume of data to be transmitted. Many traditional data aggregation schemes exploit a specific network structure (e.g., a cluster based network or forming a data aggregation tree structure). In MA-based WSNs, an MA visits the source nodes one after the other. At each source node, the MA collects sensory data and then performs data aggregation and removes any existing redundancy, depending on the data aggregation function. There is no need for a specific topology structure. However, the data aggregation and energy performance are highly dependent on the order in which the source nodes are visited, i.e., the itinerary. Consider an MA dispatched by the sink node to collect data from n source nodes. Let $l$proc be the size of the MA processing code, Shead be the size of the agent packet header, and $l0$ ma be the agent size when it is first dispatched by the, sink node. Then, we have $l0$ ma = $l$proc + Shead. Let $r \in [0, 1)$ be the reduction ratio in sensory data by agent-assisted local processing and $l$data be the size of raw data at a source node.

The reduced data payload collected by the agent at each source, which is denoted by lrd, is lrd = $(1 - r) \cdot l$data. Let lk ma be the agent size when it leaves the kth source $(1 \le k \le n)$. Since there is no data aggregation at the first source, we have $l1$ma = $l0$ma + lrd. When the agent visits the second source node, it begins to perform data aggregation to reduce the redundancy between the data collected at the current source and the data it has carried.

Let $\rho \in [0, 1]$ denote the data aggregation ratio, which is a measure of the compression performance. The MA size, after it leaves the second source node, is $l2$ma = $l0$ma + lrd + $(1 - \rho)l$rd, and so forth. For the sake of simplicity, we assume that r, ρ, and $l$data are identical for every source. 1 After visiting the kth source node.

19.4.1.1 Ability and Structure of Loses

The Loses is the development and breakthrough of managing experiment methodology and the computer simulation Technology. This system may apply in the operating decisions of enterprises, the government management decision, the army directing decision, colleges and universities education training domains and so on.

(1) The formulation of each kind of solution. Using this system's plan subsystem, the user may input the questions to be solved, the parameters of the question, and anticipated targets in the man-machine contact surface. The system searches the similar question on the database and the Internet, and finds optimal solutions according to the parameter and the goal and gives less than three simulation results. If not found or the user feels unsatisfied, it may carry on the machine plan. The system constructs solutions on the basis of question category, parameter, goal and existing logics, and the user may use system compiler to edit logics. The user may make the revision to the plan.

(2) Appraisal of each kind of solution. Using this system's appraisal subsystem, the user may draw up the plan; input the implementation environment parameter, the anticipated target in the man-machine contact surface. The system carries on the classification and the standardization to the plan (transforms system approval data format), carries on the logical reasoning simulation according to the appraisal logic under the environment parameter which the user provides and carries on the comparison with the user's goal, and it gives the measuring results. The user may edit the appraisal logics, request the system to search the similar plan on the database and the Internet and carries on the logic reasoning to compare and to give the comparison conclusion, the user makes the plan revision according to the conclusion.

(3) Choices of each kind of solution. Using this system's policy-making subsystem, the user may input the multi-wrap solution in the man-machine contact surface (to be directly carried on by the 3D modeling, the 2D modeling or text description) and the targets which the plan needs to display. The system carries on the simulation on the basis of the solution, the environment parameter proposed by the user, and carries on the quantitative analysis to the targets which are cared by the user,

gives the divided target experiment conclusion and the total performance experiment conclusion to be chosen by the user. (4) Strategy gambling of each kind of solution. Using this system's gambling subsystem, the user may carry in-line resistance or the man-machine resistance (by on-line resistance primarily). The user draws up the good solution using the man-machine contact surface according to the step input, the system carries on the simulation to the strategy according to certain logic rules and all resources, then will feedback the adjudicated statement to the user, the various users make the plan of next step according to the situation and then implement. When the gamble achieves the finishing condition, the system will send the feedback to all quarter's simulation results of victory and defeat, profit and loss. This method is suitable for military struggle gambling and the enterprise management gambling, may examines organization's strategy feasibility and the probability of victory and defeat.

19.4.2 Accuracy of Collective Fingerprints

We evaluated the quality of collective fingerprint, which is constructed with multiple instances of LFs. First, we analyzed the coverage and the accuracy of the collective fingerprint according to the number of LFs.

Distance Orders of Other Source Nodes to Test Sources

Test node	1	2	3	4
A	C	B	E	D
B	D	C	A	E
C	D	B	E	A
D	E	A	B	C
E	C	D	B	A

The coverage is the ratio of the fingerprinted grids to total grids (i.e., 80 grids) of the target area. The accuracy is the mean location error in all fingerprinted grids. The coverage and the accuracy increase in proportion to the number of collected LFs. As collected LFs. reach 40, the proposed system attains coverage of 82.5% and a location error of 5.42 m. With this experiment, we validate that the proposed system can indeed construct more accurate and larger radio map as more LFs are collected.

Algorithm 1 Pseudo Code for Collective Fingerprint Update

```
Procedure Update( LF ) //Update CF with newly uploaded LF

for all tuple T_LF (x_LF,y_LF,c_LF,S_LF) in LF
  Overlapped = false
  for all tuple T_CF (x_CF,y_CF,c_CF,S_CF) in CF
    //check overlapped grid
    if (x_LF,y_LF) = (x_CF,y_CF)
      Overlapped = true
      if c_LF > c_CF
        T_GF = T_LF
    //check for alias tuple
    else if Diff (S_LF , S_CF ) < ε_same
      if c_LF > c_CF
        T_CF = null
      else
        T_LF = null
  //copy T_LF to T_CF when no overlapping
  if Overlapped = false
    T_CF = T_LF
```

19.5 Experimental Results

This section discusses the experiment results that were conducted to evaluate the proposed system. We will first describe the platform design and experimental environments. Preliminary experiments are then explained to obtain various parameters used in a real scenario. The performance of the proposed system is compared with the ground-truth RSS fingerprint, which is constructed by a conventional site-survey method is illustrated in Fig. 19.3.

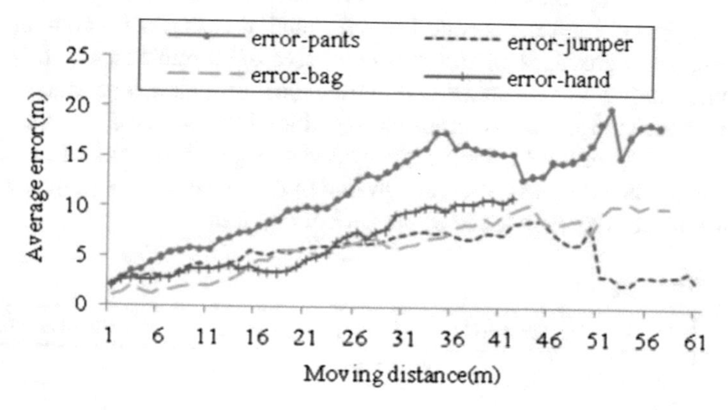

Fig. 19.3 Error prediction

19.5.1 Experimental Setup

To evaluate the proposed system in real environments, we implemented the system on the HTC Hero [11], which is an Android Smartphone equipped with both an accelerometer and a magnetometer. The experiments were conducted in an engineering building at Yonsei University. The site is 60 × 66 m and has three entrances. The number of APs detected inside the building is between 5 and 25, depending on the location of the user. To construct diverse types of LF in the experiments, we surveyed 40 participants' paths to work, as well as their phone position at each entrance, as shown in We then collected 40 LFs based on the result of the survey.

19.5.2 Localization Performance with Inertial Sensors

The current location of a mobile user is estimated by inertial sensor-based localization. The localization performance is significantly influenced by the quality of local RSS fingerprints and consequently on the quality of global RSS fingerprints. To estimate the accuracy of inertial sensor-based localization, we analyzed the average location error of all the paths. We compared the performance of two methods, described in Sect. 19.3.1, for finding heading.

19.5.2.1 Alias Filtering with RSS Difference

Such that both A and B stay at the position (x, y) and measure RSS, but their tuples TA and TB are stored at different positions because of the location error of inertial sensors-based localization, larger control effort are required. In [7], both its neighbors' states and its neighbors' with neighbors' states. By introducing more information in the second hop, the consensus convergence speed is improved. However, the trade-off for introducing the second hop information is that extra communication and the proposed consensus algorithm is composed of current states outdated states stored in memory under an undirected communication graph. The algorithm converges faster than the standard consensus algorithm if the outdated states are chosen properly. In [3], from the Laplacian eigenvalue standpoint, the author has focused on maximizing the second smallest eigenvalue of a state dependent graph Laplacian to improve the convergence speed of the network systems.

Effect of RSS Spacing

	# of AP scans per a grid	Location error (m)	Grid hit rate (%)
1 m	0.7	3.1	20.1
2 m	1.4	2.6	34.3
3 m	2.0	3.0	50.3

(continued)

(continued)

	# of AP scans per a grid	Location error (m)	Grid hit rate (%)
4 m	2.6	3.5	58.1
5 m	3.4	4.4	62.8

The control input of each agent consists of In discrete-time models, each agent updates its state by computing a weighted average of its own value with values received from their neighbors at each time step. The convergence properties of the consensus algorithms have been further studied under different assumptions on the topology connectivity and information exchange in Fig. 19.4.

19.6 Conclusions

In this paper we have developed a multi agent system that uses an inertial sensor networks, automatically storing the fingerprinting and a measure the signal range. The costly offline process of RSS map construction in conventional fingerprint-based system is removed by the use of inertial Sensors. The MAT system conducts inertial sensor-based self-localization and collect RSS fingerprints automatically the localization. Second, we successfully constructed the collective fingerprint map by a credibility-based user collaboration scheme. The proposed system is based on a passive user participation model, which collects range of signal and uploads fingerprints. For the future work, we plan to improve the proposed system in several aspects. The inertial sensor-based MAS will be enhanced to improve the quality of

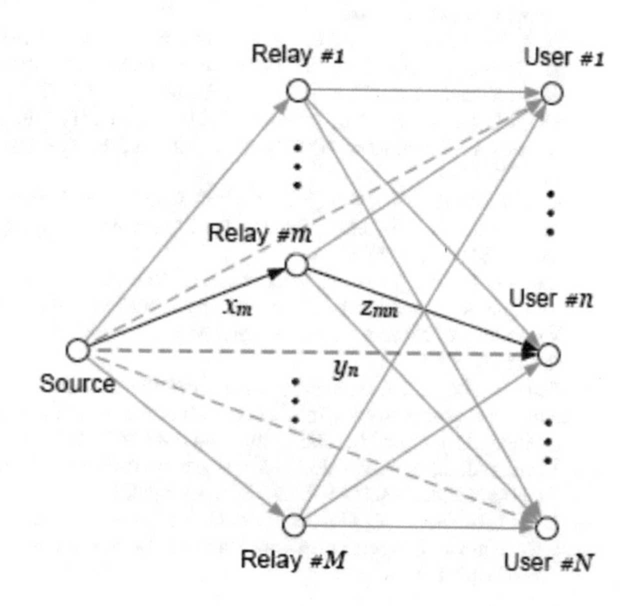

Fig. 19.4 Source-relay formation

local fingerprints. Finally, we plan to implement a MAS system by detecting the precise entrance position of the range, which was assumed to be known in the current work.

References

1. Nissanka, B.P., Anit, C., Hari, B.: The cricket location-support system. Paper presented at the 6th Annual International Conference on Mobile Computing and Networks, Boston, MA, 2000
2. Addlesee, M., Curwen, R., Hodges, S., Newman, J., Steggles, P., Ward, A., Hopper, A.: Implementing a sentient computing system. Computer **34**(8), 50–56 (2001)
3. Oliver, W., Robert, H.: Pedestrian localization for indoor environments. Paper presented at the 10th International Conference on Ubiquitous Computing, Seoul, Korea, 2008
4. Balaji, S., Golden Julie, E., Harold Robinson, Y.: Development of fuzzy based energy efficient cluster routing protocol to increase the lifetime of wireless sensor networks. Mobile Netw. Appl. **24**(2), 394–406 (2019)
5. Nattapong, S., Prashant, V.K.: On clustering RSS fingerprints for improving scalability of performance prediction of indoor positioning systems. Paper presented at the First ACM International Workshop on Mobile Entity Localization and Tracking in GPS-less Environments, San Francisco
6. Santhana Krishnan, R., Golden Julie, E., Harold Robinson, Y., Kumar, R., Son, L.H., Tuan, T.A., Long, H.V.: Modified zone based intrusion detection system for security enhancement in mobile ad-hoc networks. Wirel. Netw., 1–15 (2019)
7. Harold Robinson, Y., Golden Julie, E., Saravanan, K., Kumar, R., Son, L.H.: DRP: Dynamic routing protocol in wireless sensor networks. Wirel. Pers. Commun., 1–17, Springer (2019)
8. Wan, P.-J., Alzoubi, K.M., Frieder, O.: Distributed construction of connected dominating set in wireless ad hoc networks. Mobile Netw. Appl. **9**(2), 141–149 (2004)
9. Balaji, S., Golden Julie, E., Harold Robinson, Y., Kumar, R., Thong, P.H., Son, L.H.: Design of a security-aware routing scheme in mobile ad-hoc network using repeated game model. Comput. Stand. Inter. **66**, (2019)
10. Ding, L., Shao, Y., Li, M.: On reducing broadcast transmission cost and redundancy in ad hoc wireless networks using directional antennas. In: Proceedings of the IEEE Wireless Communications and Networking Conference (WCNC), 2008
11. Min, M., Du, H., Jia, X., Huang, C.X., Huang, S.C.-H., Wu, W.: Improving construction for connected dominating set with Steiner tree in wireless sensor networks. J. Global Optim. **35**(1), 111–119 (2006)
12. Harold Robinson, Y., Balaji, S., Golden Julie, E.: FPSOEE: Fuzzy-enabled particle swarm optimization-based energy-efficient algorithm in mobile ad-hoc networks. J. Intell. Fuzzy Syst. **36**(4), 3541–3553 (2019)
13. Barriére, L., Fraigniaud, P., Narayanan, L.: Robust position-based routing in wireless ad hoc networks with unstable transmission ranges. In: Proceedings of the ACM Fifth International Workshop on Discrete Algorithms and Methods for Mobile Computing and Communications (DIALM), 2001
14. Zeng, Y., Jia, X., He, Y.: Energy efficient distributed connected dominating sets construction in wireless sensor networks. In: Proceedings of the ACM International Conference on Wireless Communications and Mobile Computing (IWCMC), 2006
15. Yi, S., Andrzejak, A., Kondo, D.: Monetary cost-aware checkpointing and migration on amazon cloud spot instances. IEEE Trans. Serv. Comput. **5**(4), 512–524 (2012)
16. Harold Robinson, Y., Golden Julie, E., Balaji, S., Ayyasamy, A.: Energy aware clustering scheme in wireless sensor network using neuro-fuzzy approach. Wirel. Pers. Commun. **95**(2), 703–721 (2017)

17. Shen, H., Hwang, K.: Locality-preserving clustering and discovery of resources in wide-area distributed computational grids. IEEE Trans. Comput. **61**(4), 458–473 (2011)
18. Harold Robinson, Y., Rajaram, M.: Energy-aware multipath routing scheme based on particle swarm optimization in mobile ad hoc networks. Sci. World J., 1–9 (2015)
19. Cai, M., Hwang, K.: Distributed aggregation algorithms with load-balancing for scalable grid resource monitoring. In: Proceeding of the IEEE International Parallel and Distributed Processing Symposium (IPDPS), 2007
20. Cai, M., Frank, M., Szekely, P.: MAAN: a multi-attribute addressable network for grid information services. In: Proceedings of the Fourth International Workshop on Grid Computing, 2011

Chapter 20
Iron Oxide Nanoparticles in Biosensors, Imaging and Drug Delivery Applications—A Complete Tool

Aparajita Singh and Vinod Kumar

Abstract Iron Oxide Nanoparticles(IONPs) have been extensively studied and have found applications in biomedical engineering. Iron Oxide owing to its unique biocompatibility and therefore low toxicity have come up as an amazing category of nanoparticles to be used as drug delivery vehicle especially being used for cancer therapies e.g. hyperthermia. Alongside IONPs have found a great deal of application into biosensors e.g. Glucose sensor, BPA sensor, Gas sensors. In this paper we have reviewed the different biosensors based on IONPs and their applications, IONPs in imaging and drug delivery. Iron Oxide Nanoparticles can be synthesized with co-precipitation, thermal decomposition, microwave techniques to achieve optimum results and then coated with polymers and further loaded with related drug, biomolecule or a dye to obtain an optical result. These IONPs are then doped with other nanoparticles to form composites with nanostructures as Graphene or biomolecules as Chitosan are further usually dispersed over a glass or a silicon substrate to fabricate a biosensor. IONPs tend to achieve super paramagnetic property and therefore known as SPIONs that increases the application of these nanoparticles in all spectrums of biomedical application.

Keywords Iron oxide nanoparticles (IONPs) · Cytotoxicity · Imaging · Surface modification · Biosensors · Biomedical applications

20.1 Introduction

Metal oxide nanoparticles have been explored for their applications into biosensor applications for over a decade. Zinc, Tungsten, Nickel, Zirconium, Silver, Cobalt And Vanadium are some of the commonly used metal oxides being nanostructured over a silicon or glass substrate to fabricate a functional and efficient biosensor. The composites of these metal oxides have also proven to enhance the functionality based

A. Singh · V. Kumar (✉)
SRMIST-NCR Campus, Ghaziabad, India
e-mail: vinodkumar.viet@gmail.com; vkchaudhary.rs.ece@iitbhu.ac.in

A. Singh
e-mail: aparajita.singh2015@gmail.com

V. E. Balas et al. (eds.), *Internet of Things and Big Data Applications*, Intelligent Systems Reference Library 180, https://doi.org/10.1007/978-3-030-39119-5_20

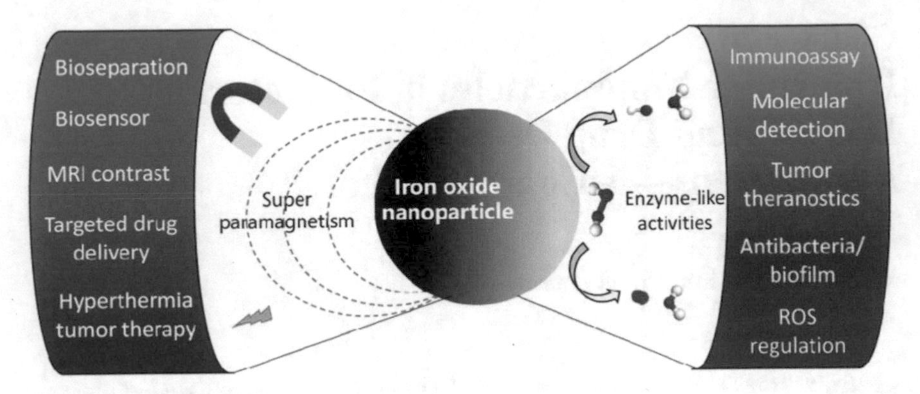

Fig. 20.1 Iron oxide nanoparticles potential applications [7]

on electrochemical or photochemical detection of these sensors. These sensors have mainly found application for glucose sensing but in the recent decade some metal oxides forming composites with graphene like nanomaterials have found applications in detection of viruses or even cancer cells when relevant antibody or biomolecule is channelized for detection. This adds up to the new possibilities and various targets for the simply synthesized metal oxides.

The reason for studying Iron Oxide as the unique metal oxide, finding best suited applications in biomedical and diagnostic sciences is due to its physiochemical properties. Being safe and biocompatible it has made it easier to work with. The low or nearly no toxicity of IONPs as reported in studies. In vivo studies have shown that Iron Oxide Nanoparticles (IONPs) after penetration into the cell stay within cell organelles and get accumulated in the organs like liver, lungs, spleen and brain after inhalation, also being capable of crossing the crucial BBB (blood-brain barrier) but, it also generates a toxic effect [1]. The capability to cross BBB makes them suitable for drug delivery and imaging diagnostics on nervous system. So along with the best output with Iron Oxide the minimal amount of toxicity can't be ignored fully (Fig. 20.1).

20.2 Iron Oxide for Drug Delivery

Steps for synthesis of Iron Oxide Nanoparticles (IONPs) for targeted drug delivery

20.2.1 Synthesis of Polymer (e.g. PVA) Coated Iron Oxide Nanoparticles

Synthesis of Polymer (e.g. PVA) Coated Iron Oxide Nanoparticles (Green synthesis) through the technique of co-precipitation has already been reported in Doxorubicin loaded PVA coated iron oxide nanoparticles for targeted drug delivery [2].

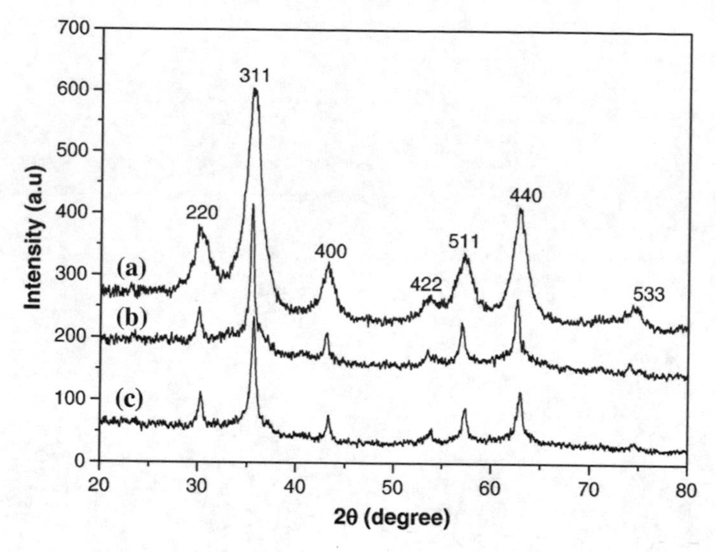

Fig. 20.2 X-ray diffraction peaks for IONPs [3]

Among all the available methods, co-precipitation method is widely used. Almost all Super-Paramagnetic Iron Oxide Nanoparticles (SPIONs) can be synthesized through this method. It is simple as well as cost-effective method. Massart's technique has gained been used with broad variety of modifications, although the main chemistry remains same (Fig. 20.2):

$$Fe^{2+} + H_2O \rightarrow Fe(OH)_2(Deprotonation)$$
$$Fe^{3+} + H_2O \rightarrow Fe(OH)_3(Deprotonation)$$
$$Fe(OH)_2 + Fe(OH)_{3\rightarrow} Fe_3O_4(Oxidation)$$

20.2.2 Surface Modifications of Iron Oxide Nanoparticles

See Fig. 20.3.

20.2.3 Drug Loading

Anti-cancer drugs such as Paclitaxel or Doxorubicin (DOX) works well with an Iron Oxide-PVA conjugate. As reported already,

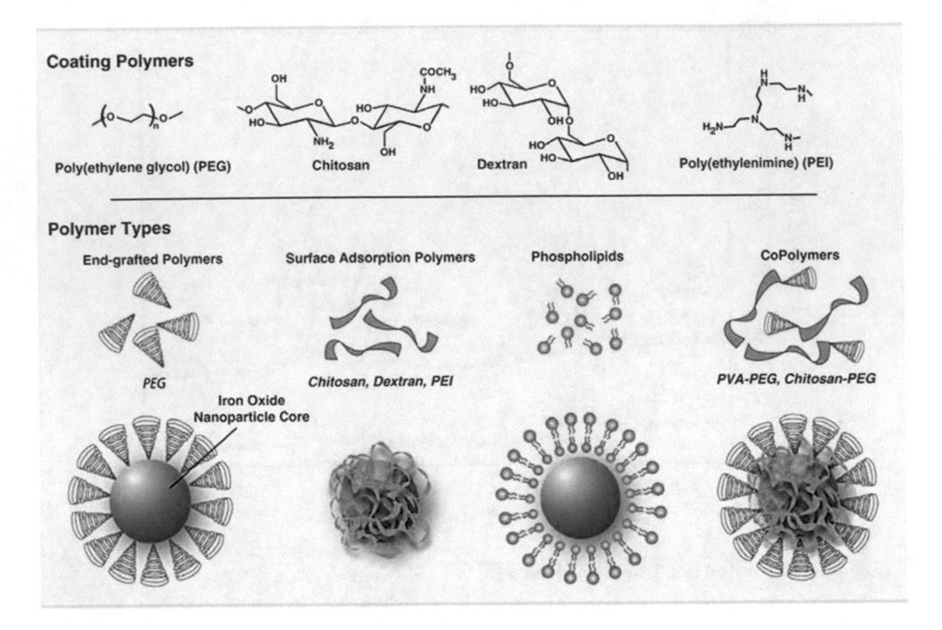

Fig. 20.3 Polymer coating of a Nanoparticle [2]

The mixture of Poly vinyl alcohol (PVA) coated Iron Oxide Nanoparticles (IONPs) in anti-cancer drug Doxorubicin (DOX) is shaken in a rotary shaker under ambient conditions to facilitate DOX uptake [3].

20.2.4 Magnetic Hyperthermia

Magnetic hyperthermia is killing the cancer cells at the site of cancer by targeted drug delivery and heat treatment. Iron Oxide Nanoparticle have magnetic properties and therefore can be directed using an external magnetic field towards the site of action.

Steps of hyperthermia-

(1) *Targeting*
(2) *Drug release*
(3) *Drug penetration* (Figs. 20.4 and 20.5).

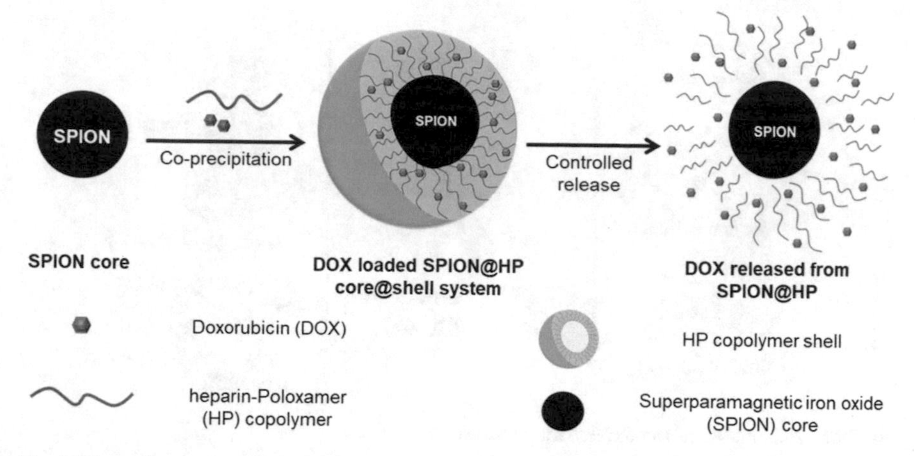

Fig. 20.4 Drug release of a core-shell structure

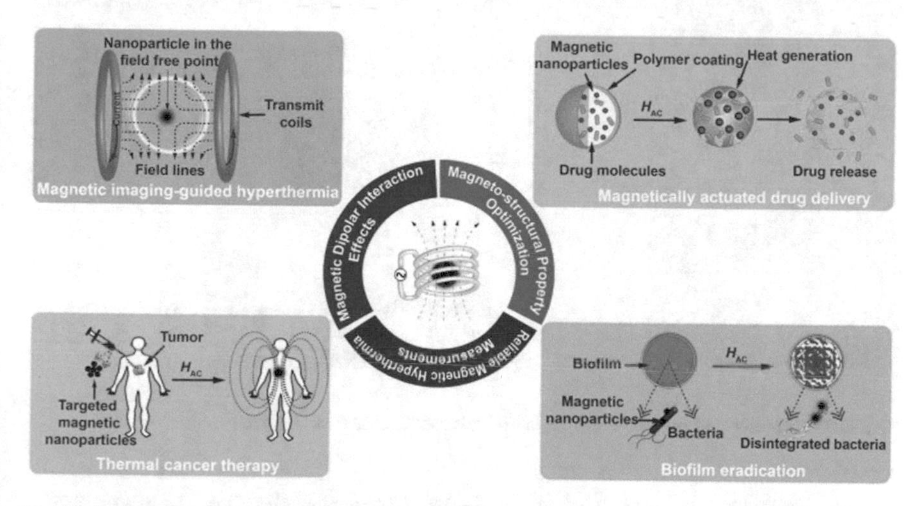

Fig. 20.5 Magnetic hyperthermia for cancer therapy [8]

20.3 Iron Oxide in Imaging

Iron Oxide Nanoparticles are being used well for cancer imaging, MRI, real-time imaging. Along with superior bio-compatibility and magnetic nanoparticles show greater sensitivity in the micro-molar or nano-molar range than, so such nanoparticles can be used as T_2 MRI contrast agents.

Magnetic Iron Oxides, Magnetite (Fe_3O_4) and Maghemite (g-Fe_2O_3), have a long-range orde- of magnetic moment.

Super-Paramagnetic Iron Oxide Nanoparticles constitute a great contrast agent that has been explored for magnetic imaging techniques (Figs. 20.6 and 20.7).

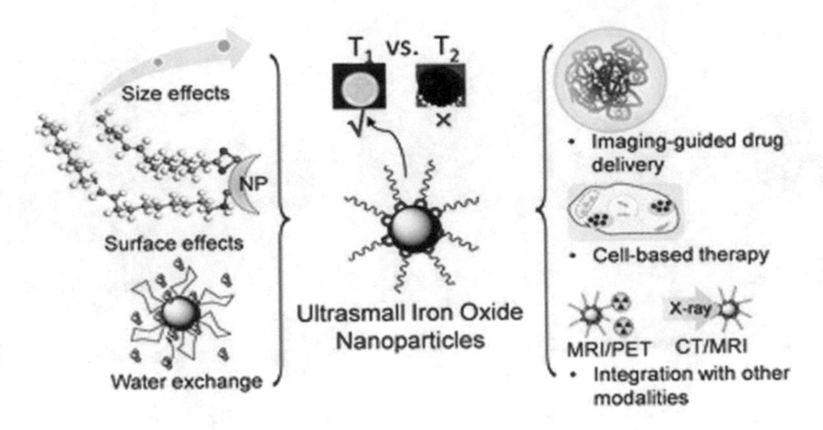

Fig. 20.6 Imaging with iron oxide nanoparticles [9]

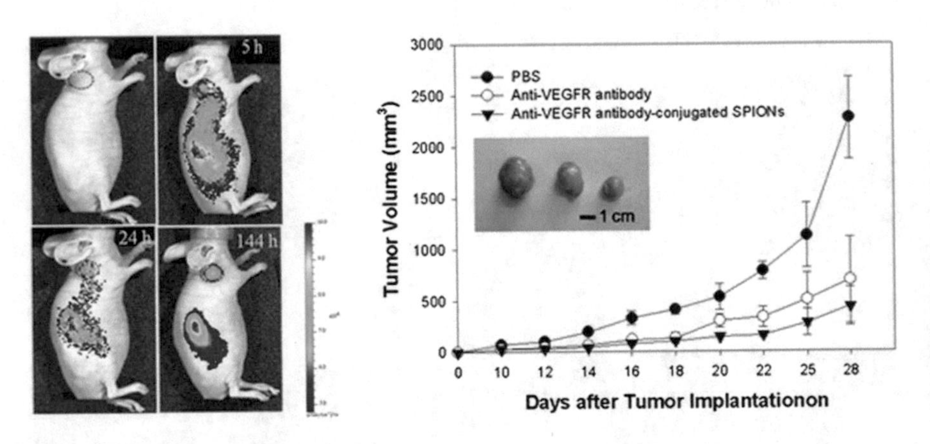

Fig. 20.7 Imaging and tumour necrosis with iron oxide nanoparticles [10]

They present a magnetic core, typically Magnetite (Fe_3O_4), Maghemite (γ-Fe_2O_3), or a mixture of both forming a crystalline structure. IONPs are also used in PET scans and nuclear imaging. It's easier to direct the magnetic nanoparticle to the site and therefore the precision is high.

Apart from imaging the same nanoparticles are coupled to deliver the drug at site and therefore makes it a more versatile process imaging and treating at the same time.

In radiology for diagnostics, the MRI contrast of biological tissue is produced due to the difference of the nuclear magnetic relaxation of water protons. The magnetic moment of IONPs changes the relaxation time of the water protons in the immediate surrounding, this produces a change of contrast. In contrast to relaxation time of protons (T_1-*relaxation time*) in the surroundings, IONPs are usually T_2 contrast agents as they increase the spin-spin relaxation time (T_2 *relaxation time*). This in turn leads to a darker signal. This signifies that IO induces the protons to move out of rotation in phase with each other so that the T_2 becomes shorter. The advantage of using IONPs is that these stay inside the intravascular space for a prolonged time period and therefore allow for longer time of image-acquisition as compared to the T_1 contrast agents.

20.4 Iron Oxide for Detection/Biosensors

Iron Oxide Nanoparticles have been used widely in biosensors. The edge comes with the high surface to volume ratio offered by the Nano-size. Nanoparticles have an amazing ability that promotes electron transfer between electrode and an active site of a biomolecule e.g. enzyme.

Immobilization of active biomolecules over the surface of magnetic nanoparticles is crucial, because of the magnetic behaviour of these bioconjugates it's likely that it improves delivery and recovery of biomolecules for its use in biomedical applications. For e.g.- Chitosan-Iron Oxide Nanoparticle composite has been used for glucose sensing (Fig. 20.8).

With biosensors finding applications in detection and diagnostics at a higher pace, IONPs based biosensors are being investigated for its potential in cancer diagnostics and therapy. On the same lines an electrochemical paper based sensor using IONPs (by Kumar et al. [4]) or aptamer-conjugated thermally cross-linked Super-Paramagnetic Iron Oxide Nanoparticles (by Yu et al. [5]) have shown the possibilities of new unexplored lines of detection and therapy of cancer (Fig. 20.9).

Conventional technology such as fluorescence microarray and electrochemical methods or DNA microarray are expensive and require high cost input for reagents

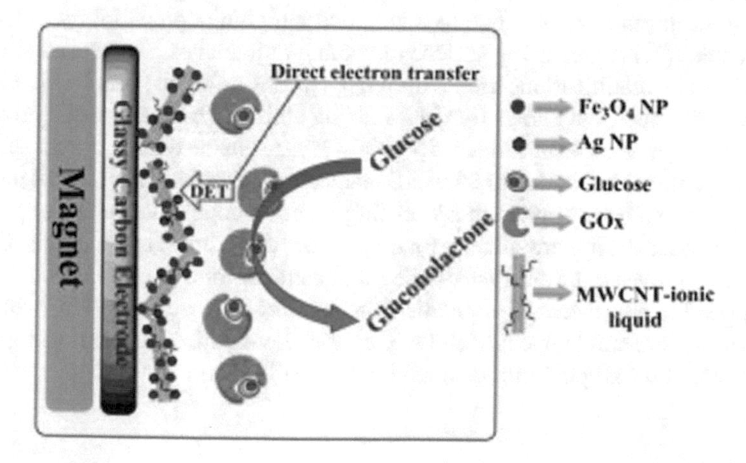

Fig. 20.8 Iron oxide based glucose sensor [11]

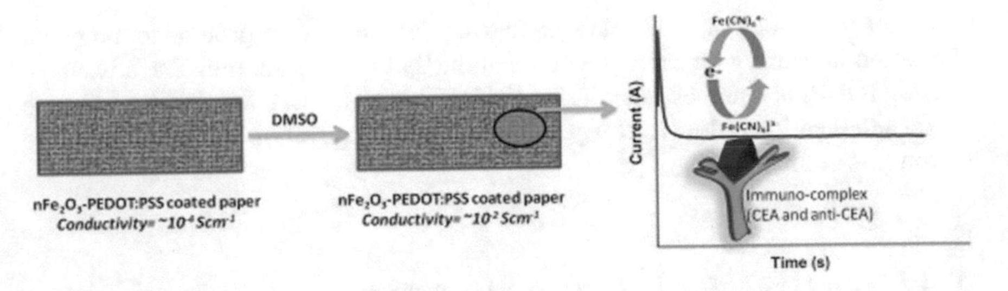

Fig. 20.9 IONPs based biosensor for cancer detection [4]

and monitoring whereas the new age sensing techniques e.g.-biosensing is inexpensive and portable e.g.-lab on a chip which makes it more useful and versatile. Also it allows real-time monitoring e.g.- a dye or a composite of IONPs and makes it suitable for clinical applications.

Moreover the composites i.e. Graphene-Iron Oxide composite and other nanocomposite based biosensor are specific to the biomarker. A bimetallic Cerium and Ferric Oxides Nanoparticles embedded in a mesoporous carbon matrix sensor (by Wang et al. [6]) is specific to ovarian cancer sensing highly specific to carbohydrate antigen 19-9. There are many ways for detection of the output (Fig. 20.10).

20.5 Conclusion

Iron Oxide nanoparticles(IONPs) have proved to be an excellent candidate for a complete therapy of cancer and others. Of course the toxicity due to the accumulation of these nanoparticles with no outlet to be removed from the body is an issue to deal with but the advantages of IONPs due to low toxicity, high biocompatibility, selectivity and size that offers an amazing surface to volume ratio makes it a great candidate to work with. The modifications tried with IONPs have also given great result, be it the core-shell structure that allows for drug-loading along with the magnetic properties to make it a great tool for guided drug delivery. It has also been a cost effective contrast agent that has added to conventional techniques as MRI/PET. Not just that, IONPs have also been applied in biosensing alone/as composites and has given an opportunity to make it more suitable for diagnostics and clinical applications. Overall IONPs have proved to be a very versatile set of nanoparticles that have been successfully applied to all spheres of biomedical engineering. The investigations with more modifications are still happening and will continue with this wonderful nanoparticle to be explored for all possibilities touching human lives in many ways.

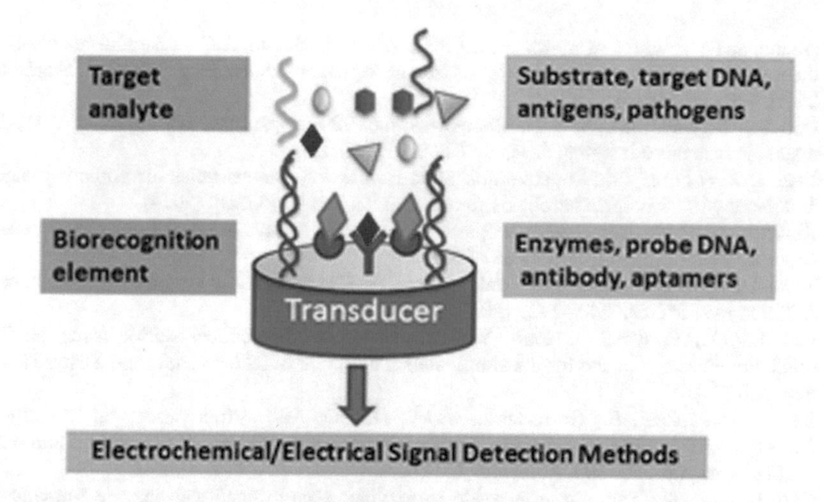

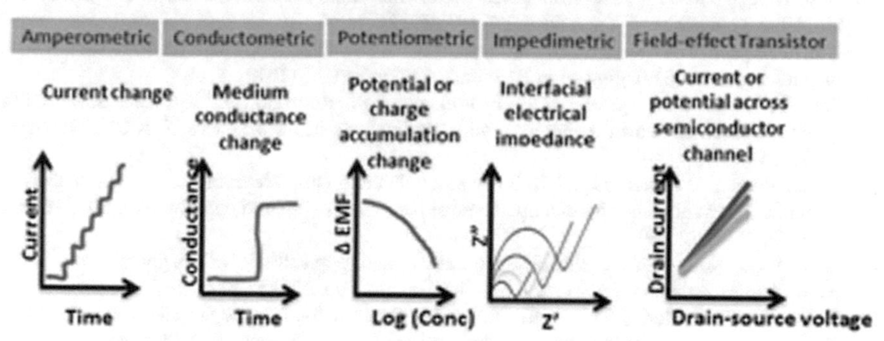

Fig. 20.10 Electrochemical signal detection process [12]

References

1. Bahadar, H., Maqbool, F., Niaz, K., Abdollahi, M.: Toxicity of nanoparticles and an overview of current experimental models. Biomed J. **20**(1), 1–11 (2016)
2. Kayal, S.., Ramanujan, R.V.: Doxorubicin loaded PVA coated iron oxide nanoparticles for targeted drug delivery. Mater. Sci. Eng. C **30**, 484–490 (2010)
3. Santra, S., Kaittanis, C., Grimm, J., Perez, J.M.: Drug loaded multifunctional iron oxide nanoparticles for combined targeted drug therapy and dual optical/magnetic resonance imaging. Small **5**(16), 1862–1868 (2009)
4. Kumar, S., et al.: Electrochemical paper based cancer biosensor using iron oxide nanoparticles decorated PEDOT: PSS. Anal. Chim. Acta **1056**, 135–145 (2019)
5. Yu, M.K., Kim, D., Lee, I.-H., So, J.-S., Jeong, Y.Y., Jon, S.: Image-guided prostate cancer therapy using aptamer-functionalized thermally cross-linked superparamagnetic iron oxide nanoparticles. Small **7**(15), 2241–2249 (2011)
6. Wang, M., Fang, S.: Bimetallic cerium and ferric oxides nanoparticles embedded within mesoporous carbon matrix: Electrochemical immunosensor for sensitive detection of carbohydrate antigen 19-9. Biosens. Bioelectron. **135**(15), 22–29 (2019)
7. Gupta, A.K., Gupta, M.: Synthesis and surface engineering of iron oxide nanoparticles for biomedical applications. Biomaterials **26**(18), 3995–4021 (2005)

8. Abenojar, E.C., Wickramsinghe, S., Basconcepcion, J., Samia, A.C.S.: Structural effects on the magnetic hyperthermia properties of iron oxide nanoparticles. Progr. Nat. Sci. Mater. Int. **26**(5), 440–448 (2016)
9. Bao, Y., Sherwood, J.A., Sun, Z.: Magnetic iron oxide nanoparticles as T1 contrast agents for magnetic resonance imaging. J. Mater. Chem. (6), (2018)
10. Jang, D.R. et al.: Hybrid Superparamagnetic Iron Oxide Nanoparticles for Tumour Imaging and therapy In Vivo. Smart Probes, Presentation number 0145, Sept (2010)
11. Kaushik, A., et al.: Iron oxide nanoparticles—chitosan composite based glucose biosensor. Biosens. Bioelectron. **24**(4), 676–683 (2008)
12. Terse-Thakoor, T., Badhulika, S., Mulchandani, A.: Graphene based biosensors for healthcare. J. Mater. Res. **32**(15), 2905–2929 (2017)
13. Lee, S., Oh, J., Kim, D., Piao, Y.: A sensitive electrochemical sensor using an iron oxide/graphene composite for the simultaneous detection of heavy metal ions. Korea **160**(1), 528–536 (2016)
14. Iosub, C.S., Olaret, E., Grumezescu, A.M., Holban, A.M., Andronescu, E.: Toxicity of nanostructures—a general approach. In: Nanostructures for Novel Therapy Micro and Nano Technologies, pp. 793–809, Romania (2017)
15. Gao, L., Fan, K., Yan, X.: Iron oxide nanozyme: a multifunctional enzyme mimetic for biomedical applications. Theranostics **7**(13) (2017)
16. Lee, N., Hyeon, T.: Designed synthesis of uniformly sized iron oxide nanoparticles for efficient magnetic resonance imaging contrast agents. Korea, RSC 7, (2012)
17. Pena-Bahamonde, J., Nguyen, H.N., Fanourakis, S.K., Rodrigues, D.F.: Recent advances in graphene-based biosensor technology with applications in life sciences. J. Nanobiotechnol. **16**(1), 75 (2018)
18. De La Franier, B., Thompson, M.: Early stage detection and screening of ovarian cancer: A research opportunity and significant challenge for biosensor technology. Biosens. Bioelectron. **135**(15), 71–81 (2019)
19. Teja, A.S., Koh, P.-Y.: Synthesis, properties, and applications of magnetic iron oxide nanoparticles. Atlanta USA, Google Scholar, **55**(1–2), pp. 22–45 (2009)
20. Cotin, G., Piant Mertz, S., Felder, D., Begin, S.: Iron oxide nanoparticles for biomedical applications: Synthesis, functionalization, and application. France, Metal Oxides, pp. 43–88 (2018)
21. Chaterjee, J., Haik, Y., Chen, C.-J.: Size dependent magnetic properties of iron oxide nanoparticles. J. Magn. Magn. Mater. **257**(1), 113–118 (2003)
22. Habib, A.H.: Evaluation of iron-cobalt/ferrite core-shell nanoparticles for cancer thermotherapy. J. Appl. Phys. **103**(7), 07A307 (2008)
23. Wahajuddin, S.A.: Superparamagnetic iron oxide nanoparticles: magnetic nanoplatforms as drug carriers. Int. J. Nanomed. **7**, 3445–3471 (2012)
24. Mahmoudi, M., Sant, S., Wang, B., Laurent, S., Sen, T.: Superparamagnetic iron oxide nanoparticles (SPIONs): Development, surface modification and applications in chemotherapy. Adv. Drug Deliv. Rev. **63**(1–2), 24–46 (2011)
25. Hilger, I., Kaiser, W.A.: Iron oxide based nanostructures for MRI and Magnetic Hyperthermia. Nanomedicine **7**, 1443–1459 (2012)

Chapter 21
An Accurate and Improved GaN HEMT Small Signal Parameter Extraction

Swati Sharma and Vinod Kumar

Abstract A highly proficient and precise algorithm is proposed for the small signal equivalent circuit parameters extraction of GaN HEMT devices. This technique incorporates direct deduction of both the intrinsic and extrinsic small signal parameters in a low frequency band. This technique is swift and goes up-to X band which fits the determined equivalent circuit very well into the S parameters. We use Cold FET and HOT FET techniques to obtain intrinsic and extrinsic parameters to demonstrate the impact of parasite element passivation, parasitic capacitances, resistances and inductances. We learn the magnitude of their effect on the efficiency of power and microwave from this stage. The validity of the suggested algorithm was carefully checked with an outstanding correlation between the parameters measured and modeled very well.

Keywords Small signal extraction · Intrinsic and extrinsic · Cold and hot FET · Parasitic capacitances

21.1 Introduction

The most suitable strong devices for high frequency and high-power microwave devices are the only one i.e. GaN HEMT device. Accurate models promote circuit design effectiveness and are vital for the designer of circuits. Much more parasites for the transistor should be considered at significantly high frequency to account for the high-frequency effects. Usually, the small-signal equivalent circuit is optimized to suitably fix the small-signal scattering parameters on the device by optimizing element values closely [1].

The purpose of this article is to present a full extraction process of all GaN HEMT equivalent-circuit elements. All extrinsic parasitic components are assessed without forward bias of the gate terminal under "cold-FET" conditions, as there is no question

S. Sharma · V. Kumar (✉)
SRM IST, Ghaziabad, UP, India
e-mail: vkchaudhary.rs.ece@iitbhu.ac.in

S. Sharma
e-mail: swati.sharma41@yahoo.com

V. E. Balas et al. (eds.), *Internet of Things and Big Data Applications*, Intelligent Systems Reference Library 180, https://doi.org/10.1007/978-3-030-39119-5_21

for a small channel resistance. A simple optimization method and a choice of well-established analytical expression could be used to accurately determine values for the intrinsic equivalent-circuit components. This research is distinguished obviously from earlier extraction methodologies by its robustness and effectiveness. The reliability and validity of the evaluated small signal modeling parameters is rigorously proved and the suggested small-signal modeling methodology is discussed.

21.2 Extraction of Small Signal Parameters

One of the most commonly used HEMT equivalent small-signal circuit is presented in the (Fig. 21.1) is best suited to GaN HEMT for RF applications.

A small signal model with 19 components is provided in this portion. The parasitic parameter includes parasitic inductances and resistances at source (Ls and Rs), drain (Ld and Rd) and gate (Lg and Rg) electrodes, parasitic capacitors between gate to drain (Cgdi), gate to source (Cgsi) and drain to source (Cdsi) as well as package-induced parasitic capacitors (Cpg and Cpd). Some kinds of optimization techniques are used to extract the value of parasitic capacitors. Under *cold-FET pinch-off* condition ($V_{ds} = 0$ V and $V_{gs} <$ Pinch-off Volt.) the circuit shown in Fig. 21.1 simplifies to a network consisting of capacitors as shown in Fig. 21.2.

The Y-parameters of this equivalent circuit can be expressed by the following equations:

$$\mathrm{Im}(Y_{11}) = j\omega(C_{pg} + C_{gsi} + C_{gs} + C_{gdi} + C_{gd}) \tag{21.1}$$

$$\mathrm{Im}(Y_{12}) = -j\omega(C_{gdi} + C_{gd}) \tag{21.2}$$

$$\mathrm{Im}(Y_{22}) = j\omega(C_{pd} + C_{dsi} + C_{ds} + C_{gdi} + C_{gd}) \tag{21.3}$$

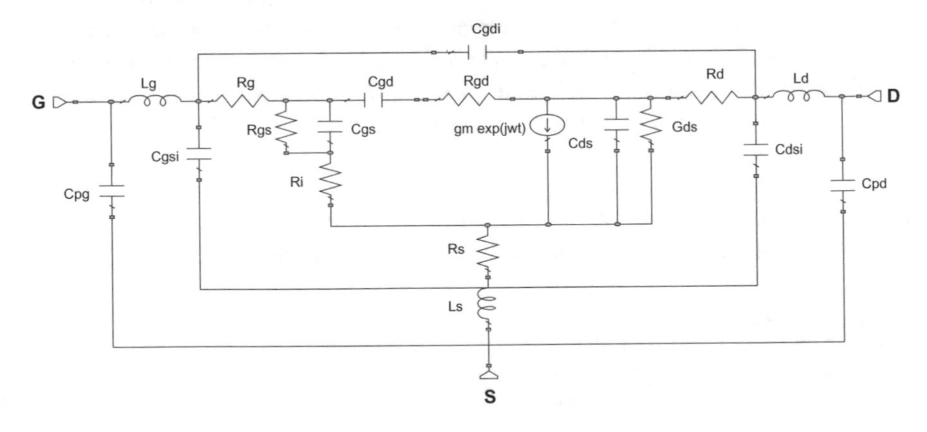

Fig. 21.1 19 elements GaN HEMT small-signal equivalent circuit model

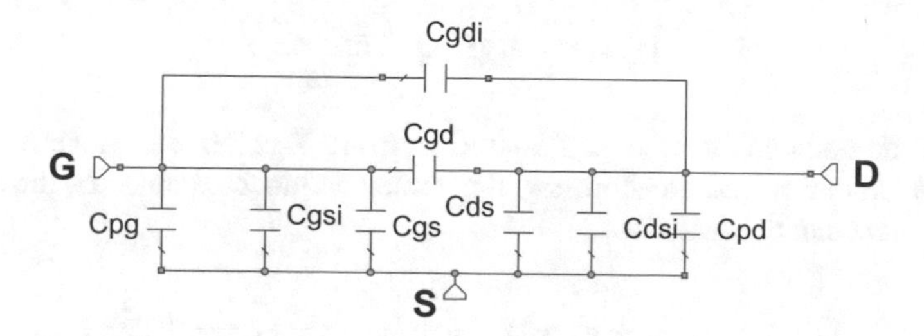

Fig. 21.2 Equiv. circuit of Fig. 21.1 under cold-FET pinch-off condition at low freq

where, $\omega = 2\pi f$ is the angular frequency.

Now for a certain geometry of the device, the value of Cpd, Cpg, Cgsi, Cgdi, Cdsi can be obtained using some extraction method. The extracted values for gate widths of 100 μm and gate lengths of 0.25 μm are:

In fact, the small signal model (Fig. 21.1) shown above can be simplified to the capacitive network shown in figure (Fig. 21.2) under cold pinch off biased only at low frequencies (<10 GHz). At comparatively medium and high frequency ranges (>10 GHz), the T-network is represented by intrinsic transistor of the pinch-off model i.e. the equivalent pi (π)-network is transformed into T-network as shown in Fig. 21.3:

The consecutive equations present the Cg, Cd and Cs for the above shown T-network [2].

$$C_d = C_{ds} + C_{gd} + \frac{CdsCgd}{Cgs} \tag{21.4}$$

$$C_s = C_{gs} + C_{ds} + \frac{CgsCds}{Cgd} \tag{21.5}$$

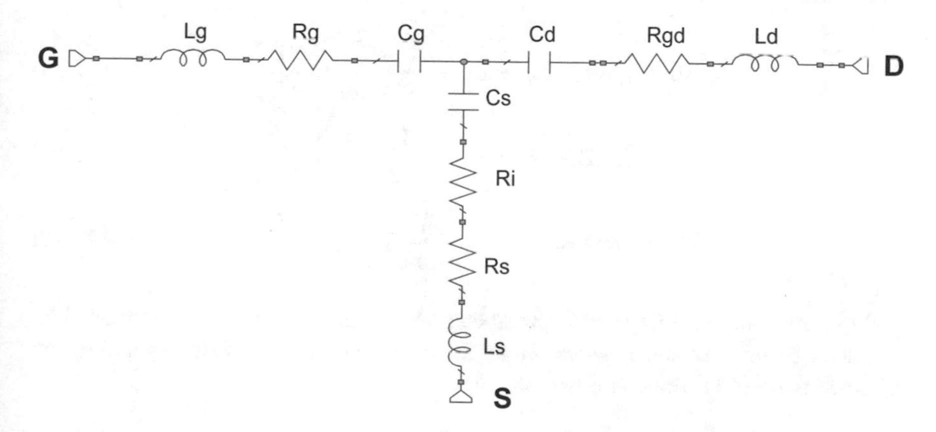

Fig. 21.3 Equiv. circuit of Fig. 21.1 under cold-FET pinch-off condition at high freq

$$C_g = C_{gs} + C_{gd} + \frac{CgsCgd}{Cds} \quad (21.6)$$

After de-embedding C_{pg}, C_{pd} and C_{gd} from Y-parameter under *cold-FET pinch-off* condition and transforming Y-parameters into Z-parameters, the stripped Z-parameters can be written as:

$$Z_{11} = R_g + R_s + R_i + j\omega(L_g + L_s) + \frac{1}{j\omega}\left(\frac{1}{Cg} + \frac{1}{Cs}\right) \quad (21.7)$$

$$Z_{12} = Z_{21} = R_s + R_i + j\omega L_s + \frac{1}{j\omega}\left(\frac{1}{Cs}\right) \quad (21.8)$$

$$Z_{22} = R_d + R_s + R_i + R_{gd} + j\omega(L_d + L_s) + \frac{1}{j\omega}\left(\frac{1}{Cd} + \frac{1}{Cs}\right) \quad (21.9)$$

It is easy to deduce the parasitic inductances Lg, Ld and Ls by the slopes of straight lines interpolating the experimental information of the imaginary components of the Z-parameters multiplied by the angular frequency ω versus ω^2, as follows:

$$\mathrm{Im}(\omega Z_{11}) = \omega^2(L_g + L_s) - \left(\frac{1}{Cg} + \frac{1}{Cs}\right) \quad (21.10)$$

$$\mathrm{Im}(\omega Z_{12}) = \omega^2(L_s) - \left(\frac{1}{Cs}\right) \quad (21.11)$$

$$\mathrm{Im}(\omega Z_{22}) = \omega^2(L_d + L_s) - \left(\frac{1}{Cd} + \frac{1}{Cs}\right) \quad (21.12)$$

Similarly, the values of Cg, Cs and Cd can be achieved from the Z-parameter plot multiplied by (1/ω) versus. (1/ω2) and calculated from the slopes of the straight lines in the following terms [3]:

$$\mathrm{Im}(Z_{11}/\omega) = L_g + L_s - \frac{1}{\omega 2}\left(\frac{1}{Cg} + \frac{1}{Cs}\right) \quad (21.13)$$

$$\mathrm{Im}(Z_{12}/\omega) = L_s - \frac{1}{\omega 2}\left(\frac{1}{Cs}\right) \quad (21.14)$$

$$\mathrm{Im}(Z_{22}/\omega) = L_d + L_s - \frac{1}{\omega 2}\left(\frac{1}{Cd} + \frac{1}{Cs}\right) \quad (21.15)$$

By taking only real parts of Z-parameter after multiplying the de-embedded Z-parameter by ω^2 resistance values can be extracted. The values of Rd, Rg and R_s are extracted from the following equations [4]:

$$\omega^2\mathrm{Re}[Z_{11}] = \omega^2(R_g + R_s) \quad (21.16)$$

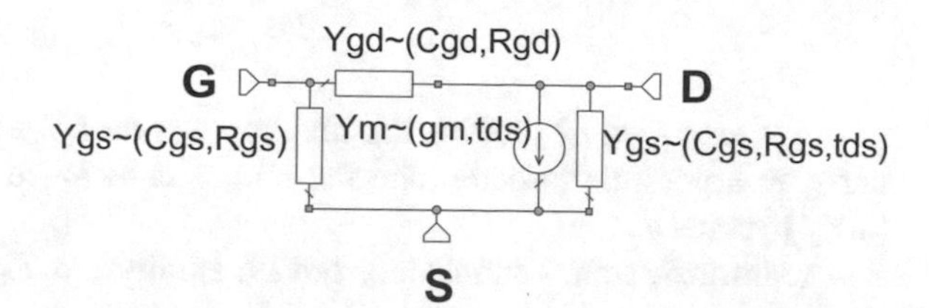

Fig. 21.4 Generalized equivalent intrinsic topology

$$\omega^2 \mathrm{Re}[Z_{22}] = \omega^2 (R_s) \tag{21.17}$$

$$\omega^2 \mathrm{Re}[Z_{22}] = \omega^2 (R_d + R_s) \tag{21.18}$$

The extrinsic parameters must be removed by extracting intrinsic parameters from the S-parameter measurements. The method of extraction of intrinsic parameters is carried out under hot-FET conditions (Vds = 28 V and Vgs = −2.5 V) [4]. The intrinsic parameters depend on bias but are independent of frequency. A generalized intrinsic topology is given in Fig. 21.4.

According to the following equation, the intrinsic gate-source branch, Ygs is calculated [2]:

$$Y_{gs} = Y_{11} + Y_{12} = \frac{G_{gsf} + j\omega C_{gs}}{1 + R_i G_{gsf} + j\omega R_i C_{gs}} \tag{21.19}$$

A variable, D is defined as [4]

$$D = \frac{\left|Y_{gs}\right|^2}{Im\left[Y_{gs}\right]} = \frac{G_{gsf}^2}{\omega C_{gs}} + \omega C_{gs} \tag{21.20}$$

The slope of curve ωD versus ω^2 plot yields the gate to source capacitance, C_{gs}. Again, by redefining D as

$$D = \frac{Y_{gs}}{Im[Y_{11}]} = \frac{G_{gsf}}{\omega C_{gs}}(1 + G_{gsf}) + \omega R_i C_{gs} - j \tag{21.21}$$

So, Ri can be extracted by plotting the Re [ωD] versus ω^2. The intrinsic gate to drain admittance, Y_{gd} can be calculated according to the following equation:

$$Y_{gd} = -Y_{12} = \frac{G_{gdf} + j\omega C_{gd}}{1 + R_{gd} G_{gdf} + j\omega R_{gd} C_{gd}} \tag{21.22}$$

It is possible to obtain Ggdf, Cgd and Rgd using the same strategy to remove Ggsf, Ri and Cgs. The intrinsic drain-source branch Yds' admittance can be articulated as the subsequent expression:

$$Y_{ds} = Y_{22} + Y_{12} = G_{ds} + j\,\omega C_{ds} \tag{21.23}$$

C_{ds} is extracted by plotting the imaginary part of Y_{ds} (Im [Y_{ds}]) against ω. Similarly, it is possible to determine the trans conductance G_{ds} from the curve of Re [ωY_{ds}] versus ω.

The intrinsic trans conductance branch admittance Y_m can be defined as,

$$Y_m = Y_{21} - Y_{12} = \frac{G_m e^{-j\omega\tau}}{1 + R_i G_{gsf} + j\omega C_{gs}} \tag{21.24}$$

Again, redefining D as,

$$\left| D = \frac{Y_{gs}}{Y_m} \right|^2 = \left(\frac{G_{gsf}}{G_m}\right)^2 + \left(\frac{C_{gs}}{G_m}\right)^2 \omega^2 \tag{21.25}$$

The slope of D versus ω^2 graph yields Gm. Trans conductance delay, τ is calculated by redefining variable D as [4],

$$D = \left(G_{gsf} + j\,\omega\,C_{gs}\right)\left(\frac{Y_m}{Y_{gs}}\right) = G_m e^{-j\omega\tau} \tag{21.26}$$

Trans-conductance delay is thus obtained by plotting the phase against ω. Extraction of intrinsic parameters using the suggested procedure provides the intrinsic variables with precise values [6]. The conventional techniques used to acquire parasitic gate and drain pad capacitances (Cpg and Cpd) from "cold-FET", $V_{ds} = 0$ V, using S-parameter measurements below pinch-off biasing conditions ($V_{gs} < V_p$). For our measurements, S-parameters are measured at $V_{gs} = -4$ V (Fig. 21.5).

Because of the expansion of the depletion layer on both sides of the gate, the star-connected network of gate capacitance elements C_b exploits the symmetrical behavior of the depletion capacitance below the gate [7]. Under open channel circumstances between the gate drain regions, a distributed RC network can be imagined as the following figure (Fig. 21.6).

Here, ΔR_c represents the distributed effective channel resistance between sources to drain terminals respectively. While ΔRdy and ΔC_g are considered, on account for the equal impedance beneath the Schottky barrier. Due to this reality, for any biasing

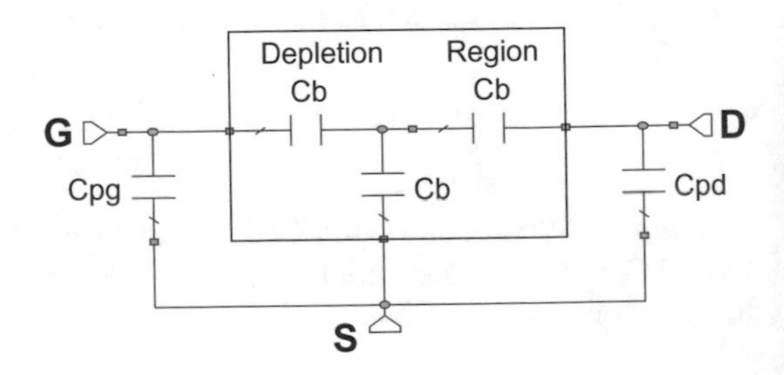

Fig. 21.5 Small-signal equivalent circuit of a FET at *pinch-off* condition

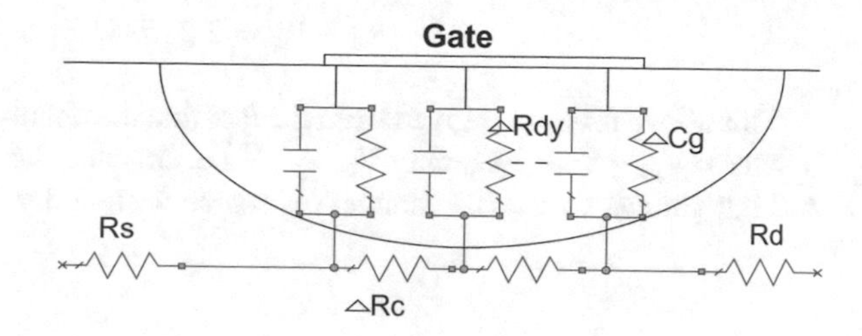

Fig. 21.6 Distributed RC network under the gate under open channel biasing condition

situations the impedance parameters can be written as,

$$z_{11} = \frac{R_C}{3} + z_{dy} \tag{21.27}$$

$$z_{12} = z_{21} = \frac{R_C}{2} \tag{21.28}$$

$$z_{22} = R_c \tag{21.29}$$

where, R_c is the channel resistance below the gate and z_{dy} represents the Schottky barrier equivalent impedance [4]. z_{dy} can be written as:

$$z_{dy} = \frac{R_{dy}}{1 + j\omega C_g R_{dy}} \text{ with } Rdy = \frac{nkT}{qI_g} \tag{21.30}$$

where k is the Boltzmann constant, n is the ideality factor, T is the room temperature, C_g is the gate capacitance and I_g is the gate current. When the gate current increases, Cg increases and Rdy decreases. Neglecting the effect of pad capacitances and gate capacitance, the extrinsic Z-parameter including the parasitic resistances and inductances can be written as [5],

$$Z_{11} = R_s + R_g + \frac{R_C}{3} + \frac{nkT}{qI_g} + j\omega(L_s + L_g) \tag{21.31}$$

$$Z_{12} = Z_{21} = R_s + \frac{R_C}{2} + j\,\omega\,L_s \tag{21.32}$$

$$Z_{22} = R_s + R_d + R_c + j\omega(L_s + L_d) \tag{21.33}$$

These expressions indicate that, while, the imaginary part of the Z-parameters increases linearly and the real part is frequency autonomous. It is to be noted that the real part of Z-parameters increases with decreasing I_g. Now from these three above expressions, the parasitic inductances will be extracted. But parasitic resistances cannot be extracted as there are five unknown including channel resistance (R_c) and I_g.

Here the relation between R_s and R_d is given by the relation as following:

$$R_d - R_s = 2.2159418 \tag{34}$$

This above relation is given from the fact that the channel resistance, R_c is proportional to $(V_{gs} - V_{th})^{-1}$ keeping $V_{ds} = 0$ V i.e. shorting the drain and source terminal. At high voltage of V_{gs} the channel resistance will tend to zero.

21.3 Results and Discussion

To verify the reliability of the extraction procedure, a 2 × 100 μm GaN HEMT, the S-parameters are measured in the range 100–20 GHz. The whole extraction procedure is implemented as C++ program. Then the extracted initial values are optimized by using an optimization algorithm of AWR MO software. In general, the allowable range is within 10% of the initial value. The optimized values for the model elements of a 2 × 100 μm GaN HEMT are listed in Table 21.1.

Simulated modeled and measured S parameters matches excellently at the bias point Vgs = −4 V and Vds = 28 V as shown in Fig. 21.7. The parameters of small signal electrical equivalent circuit model of 2 × 100 um gate periphery and 0.25 um gate length have been extracted, optimized and simulated S-parameters of the small signal model have been found to be excellently matching with measured S-parameters of the GaN HEMT (error <5%).This accordingly validates the accuracy of the proposed model and the extraction method. In order to demonstrate the importance of the two parameters Ggsf and Ggdf to GaN HEMT, a comparison is performed between the measured and simulated S-parameters based on the conventional circuit model without Ggsf and Ggdf. As is shown in Fig. 21.7, the measured and modeled S-parameters do not fit well even at low frequencies. Neglecting the Ggsf and Ggdf may impart a profound effect onto Rgd and Ri at lower frequencies. Therefore, Ggdf and Ggsf cannot be neglected for GaN HEMT because they are used in characterizing the current conduction of the gate diode applicable for not only large-signal analysis but also small-signal extraction. A legitimate large signal description of the device may be received by enforcing the bias dependences of the intrinsic elements into a circuit simulator.

Table 21.1 Extracted values of parasitic capacitors

C_{pd}	0.012838 pF
C_{pg}	0.008435 pF
C_{gsi}	0.193435 pF
C_{gdi}	0.115238 pF
C_{dsi}	0.009828 pF

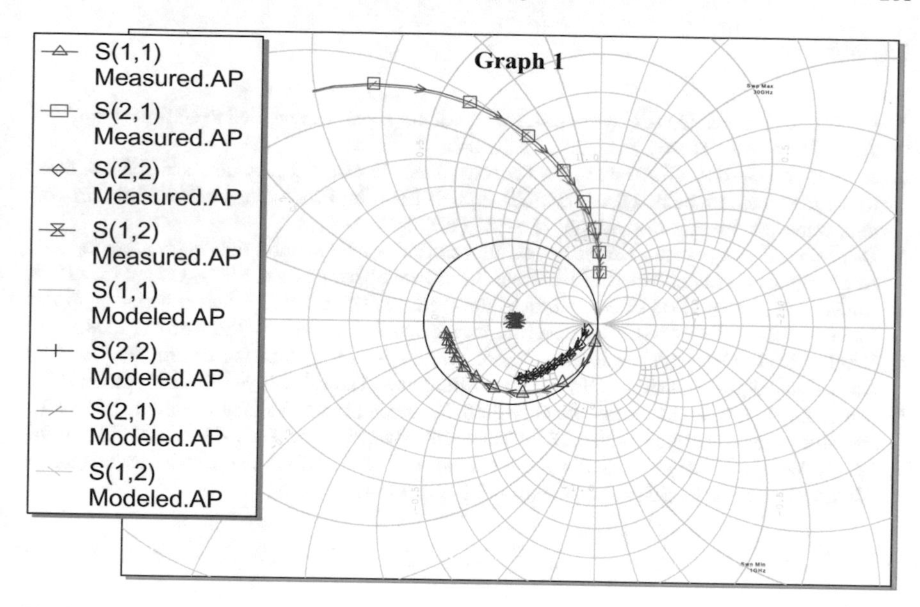

Fig. 21.7 Measured S parameters of 2*100 μ GaN HEMT and modeled S parameters

21.4 Conclusion

This document proposes a small-signal distributed 20-element model and develops easy but effective techniques for direct extraction for high frequency applications of the GaN HEMTs. Together with "cold" circumstances under different frequency range, i.e., extrinsic components are extracted from two groups S-parameters. Vds = Vgs >6 GHz and Vds <Vth; Vds = 0 and Vgs = 0 more than 6 GHz. The intrinsic components are obtained and developed over a wide frequency range from simple linear data, considering the frequency-dependent impact. All parasitic were assessed using conditional "cold FET" methods using a modified channel model that reflects the important capacitance of the drain source observed under these circumstances. In addition, it is prevented by forward biasing of the gate terminal to suppress the channel resistance. Simple terms are used to assess these parasitic quantities and thereby enhance the effectiveness of the systematic optimization method. To assess its behavior across its bias plane, a straightforward optimization method has been defined. The intrinsic elements obtained were equally independent of the frequency over the entire frequency spectrum. The validity of all extracted values was verified with an excellent correlation of measured and modelled parameters to 20 GHz.

References

1. Jarndal, A., Kompa, G.: A new small-signal modelling approach applied to GaN devices. IEEE Trans. Microwave Theory Tech. **53**(11), 3440–3448 (2005)
2. Tayrani, R., Gerber, J.E., Daniel, T., Pengelly, R.S., Rhode, U.L.: A new and reliable direct parasitic extraction method for MESFETs and HEMTs. In: Proceedings of the 23rd European microwave conference, Madrid, pp. 451–453 (1993)
3. Dambrine, G., Cappy, A., Heliodore, F., Playez, E.: A new method for determining the FET small-signal equivalent circuit. IEEE Trans. Microwave Theory Tech. **36**(7), 1151–1159 (1998)
4. Brady, R.G., Oxley, C.H., Brazil, T.J.: An improved small signal parameter extraction algorithm for GaN HEMT devices. IEEE Trans. Microwave Theor. Tech. **56**(7), 1535–1544 (2008)
5. Dambrine, G., Cappy, A., Heliodore, F., Playez, E.: A new method for determining the FET small signal equivalent circuit. IEEE Trans. Microwave Theor. Tech. **36**(7), 1151–1159 (1998)
6. Rorsman, N., Garcia, M., Karlsson, C., Zirath, H.: Accurate small signal modelling of HFET's for milli meter-wave applications. IEEE Trans. Microwave Theor. Tech. **44**(3), 432–437 (1996)
7. Lu, J., Wang, Y., Ma, L., Yu, Z.: A new small signal modelling and extraction method in AlGaN/GaN HEMTs. Solis State Electron. **52**(28), 115–120 (2008)

Chapter 22
Periodical Development of Digital Watermarking Technique

R. Vasantha Lakshmi, S. Shyam Mohana, N. Radha and Durgesh Nandan

Abstract Digital watermarking technique used to hide the information. Wide uses of internet that has increases the access of digital data like image, audio and video. There is a chance to theft the data without permission of the owner of data. To protect the copyrights of information digital watermarking is required. In this paper, the brief review on digital watermarking techniques is evaluated and described. The watermarking is produced since the image contented and could be preserved as an arithmetical impression of finger print of the image. By way of a change based on procedure is cast-off to encrypt the evidence in the histogram area that projected watermarking is vigorous sufficient in contradiction of any deprivation, equitation and occurrence.

Keywords Digital watermarking · Spatial domain · Frequency domain · Image processing · Copyright protection

22.1 Introduction

There are a lot of chance to pirate and copy the original data of the owners which were uploaded on the internet. The data can be easily downloaded and modified and commercialized on the internet. To overcome this problem, it is used digital watermarking [1]. The digital watermarking has two techniques, embedded water marking and extracted water marking technique. These techniques are performed by using the embedded and extraction process. The embedded process is inserting the watermark data into original data whereas in extracted procedure the digital watermark is removed from original data of the owner [2]. To get the effective watermark

R. Vasantha Lakshmi · S. Shyam Mohana · N. Radha
Department of ECE, Aditya Engineering College, Surampalem, East Godavari, India
e-mail: vassu1910@gmail.com

S. Shyam Mohana
e-mail: sappamohana@gmail.com

D. Nandan (✉)
Accendere Knowledge Management Services Pvt. Ltd., CL Educate Ltd., New Delhi, India
e-mail: durgeshnandano51@gmail.com

V. E. Balas et al. (eds.), *Internet of Things and Big Data Applications*, Intelligent Systems Reference Library 180, https://doi.org/10.1007/978-3-030-39119-5_22

process they are some properties [3]: (i) imperceptibility: observed that the feature of original or unique image should not be despoiled by using of watermarking (ii) robustness: It is impossible to remove the embedded digital data by unauthorized person (iii) unambiguous: recovery of the watermarking should definitely identify the owner. (iv) Loyalty: The take out watermarking should exclusively identified the original owner in the image (v) computational cost: computation cost measured of the computing resources and it should be low as possible (vi) Interoperability: Three-dimensional image combination and conversation has most important issue and deal with Interoperability (vii) unobtrusive: to recognize the watermarking gesture fullness should be comparatively minor connected to the average amplitude of contented. Digital watermarking means to insert watermark into the data to protect the data from the copyright and recognize manipulation [4]. The most requirements of digital watermarking techniques are fidelity, robustness, security, integrity, capacity and imperceptibility. The main applications of digital watermarking are copyright protection, copy control and enhance coding [5–7]. In this paper learned about the digital watermarking. The brief introduction of digital watermarking is described in Sect. 22.1. The literature has been discussed in Sect. 22.2. In Sect. 22.3 discussed the classification and comparison of watermarking techniques. In Sect. 22.4 discussed different spatial domain techniques and its comparison. In Sect. 22.5 discussed robust digital watermarking system by using feature extraction. Finally, the conclusions are given in Sect. 22.6.

22.2 Literature Review

Logo, trademark or an image are embedded in multimedia objects to prevent it from theft. Watermarking can be done by using appropriate algorithms which plays an important role in watermarking. Digital watermarking has namely spatial [8] and transform domain [8–12]. Transform domain has better robustness and imperceptibility when compare with spatial domain [12]. Digital watermarking is classified into [12] blind, visible and semi-blind [13] techniques. There are many algorithms are available in the literature for image, video and audio watermarking. Image watermarking is used to embed the original data into digital image. The authors in [1–5] embedded watermark in images. But these algorithms suffer from high computational time and the reduction in quality of the image. In audio watermark, we use audio system for watermark like MP3. In video watermarking, the original data is embedded in video stream, and it requires real time extraction [14]. The authors in [12, 13, 15, 16] watermarking is embedded in all frames of video sequences. But these algorithms suffer from high computational time and the reduction in quality of the video to overcome the drawback of image watermarking [12, 13, 15, 16]. Tabassum and Islam introduced identical frame extraction based digital video watermarking [17]. But in this method, each video shot is embedded with watermark. Using SVD and DWT, Agilandeeswari and Ganesan developed a novel algorithm [18] on video watermarking. In this algorithm, the watermark is embedded in non-motion frames

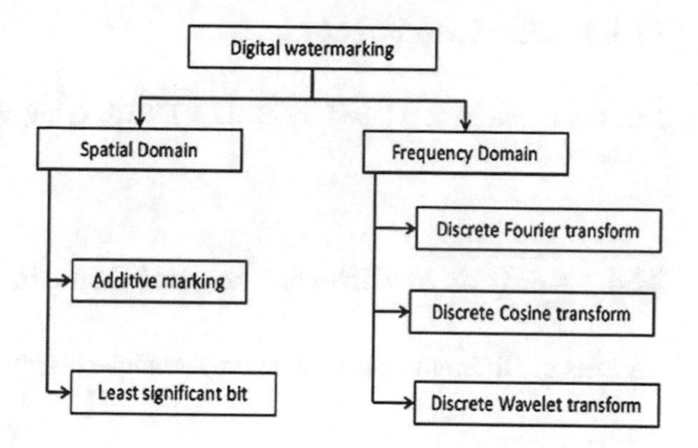

Fig. 22.1 Classification of digital water marking

extracted using histogram difference-based scene change detection. However, in this algorithm considering the small number of video frames for watermarking the watermarked video quality is reduced. Shot segmentation and block classification-based video watermarking are presented by authors in [19]. But this algorithm suffers from high computational complexity (Fig. 22.1).

22.3 Evaluation of Existing Digital Watermarking Process and Comparison

22.3.1 Spatial Domain

In this domain watermark is embedded directly into the original image. The spatial domain is manipulating an image representing an object in space to enhance the image for a given application [20]. Types of a spatial domain methods are

22.3.2 Additive Watermarking

It is the direct method used for embedding the watermark in the spatial domain [21].

22.3.3 Least Significant Bit

In this method, we insert the watermark in the LSB of pixels. It is easy to implement and very hard against the attacks [22]. This method is not used for practical applications [23].

22.3.4 Patchwork Technique

This technique is developed by Bender ETAli, using pseudo-random selection of patches [24].

22.4 Analysis of Different Spatial Domain Techniques

Analysis of Different Spatial Domain Techniques is given in Table 22.1.

22.4.1 Frequency Domain

In this domain, the watermark is embedded in the frequency coefficients of the image. This technique is widely used due to its efficiency [24]. The watermarking detection shown in Fig. 22.2 that described the feature extraction of watermarking and differentiate the original and detected image [13].

Table 22.1 Comparative analysis of different spatial domain techniques

S. No.	Technique	Advantages	Disadvantages
1	Additive watermarking	Implementation is easy. High perceptual transparency Low degradation of image processing	Subtle to noise. Susceptible to cropping and scaling. Less robustness to attacks
2	LSB	Low degradation of image processing	High perceptual transparency
3	Patchwork	High level of robustness against attacks	The small amount of information can be hidden

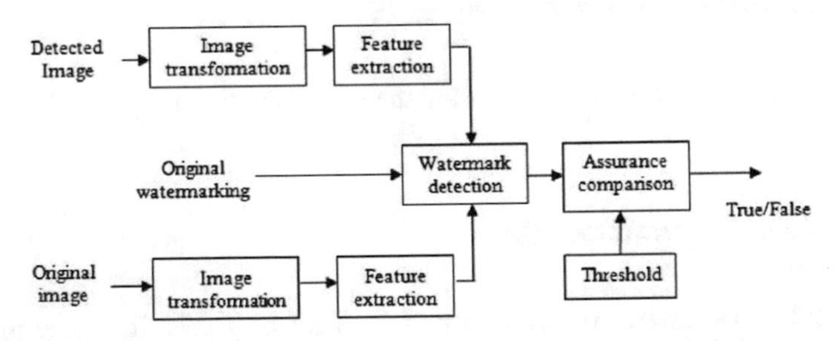

Fig. 22.2 Block diagram representation of digital water marking system by using feature extraction

Table 22.2 Comparative analysis of different frequency domain techniques

S. No.	Technique	Advantages	Disadvantages	Applications
1	DCT	More robust against attacks. Difficult to remove the watermark by any attacks because of embedding [2]	Susceptible to cropping, scaling It destroys the in-variance property of the system Difficult to overcome from geometric distortions [1]	Copyright protection Copy control
2	DFT	This technique is used to recover from geometric distortions	Complex implementation High computational cost	Broadcasting monitoring of video sequences
3	DWT	Vulnerable to cropping and scaling. It allows good localization for both time and spatial domain	More complexity. Compression time is more. Cost of computing is high difficult to overcome from geometric distortions [3]	Video authentication fingerprinting

22.5 Robust Digital Watermarking System by Using Feature Extraction

22.5.1 DCT

Discrete cosines transform Watermarking techniques are more robust compared to spatial domain techniques [14, 25]. But these methods introduce artifacts. Analysis of Different Frequency Domain Techniques is given in Table 22.2.

22.5.2 Analysis of Different Frequency Domain Techniques

22.5.3 DFT

Discrete Fourier transform has robustness against geometric attacks. It transforms a pixel into its frequency component [24]. But the computational complexity of these methods is high.

Table 22.3 Comparative analysis of digital water marking techniques

S. No.	Parameters	Spatial domain	Frequency domain
1	Computational cost	High [20]	Low
2	Robustness	Fragile	More robust
3	Perceptual quality	High	Low
4	Capacity	High	Low
5	Applications	Authentication	Copyrights

22.5.4 DWT

Discrete wavelet transform is a contemporary technique for eliminating the drawbacks of DCT and DFT. In this technique, Wavelet filters are used to transform the image [18].

22.5.5 Comparison of Digital Watermarking Techniques

It can be observed from Table 22.3 that the frequency domain water marking techniques are more efficient than spatial domain watermarking techniques.

22.6 Conclusion

In this paper, the brief review on digital water marking techniques is described. Spatial domain techniques are easy to implement, but susceptible to noise. Frequency domain techniques are not susceptible to noise and also robust to attacks. We conclude that performance of Frequency domain techniques are superior compared to others. Other things is that, the watermarking is an exclusive to specific image, in addition will be terminate entirely even if minor variation, that is a stuff to in contradiction of hacking. As a final point, meanwhile we are by means of conversion-based procedure to encrypt the evidence in the histogram field, therefore this information is strong sufficient compared to density and common place image processing. However, to store the autograph worth determined, even if the sequential quantity, in instruction to recognize the precision of the watermark.

References

1. Cox, I., Miller, M., Bloom, J., Fridrich, J., Kalker, T.: Digital watermarking and steganography. Morgan Kaufmann, San Francisco, CA (2008)
2. Petitcolas, F.A., Katzenbeisser, S.: Information Hiding Techniques for Steganography and Digital Watermarking (Artech House Computer Security Series). Artech House (2000)
3. Tao, H., Chongmin, L., Zain, J.M., Abdalla, A.N.: Robust image watermarking theories and techniques: A review. J. Appl. Res. Technol. **12**(1), 122–138 (2014)
4. Hartung, F., Girod, B.: Watermarking of uncompressed and compressed video. Sig. Process. **66**(3), 283–301 (1998)
5. Cox, I.J., Miller, M.L., Bloom, J.A., Honsinger, C.: Digital watermarking, vol 53. Springer, Berlin (2002)
6. Mistry, D.: Comparison of digital water marking methods. Int. J. Comput. Sci. Eng. **2**(09), 2905–2909 (2010)
7. Tripathi, S., Jain, R., Gayatri, V.: Novel DCT and DWT based watermarking techniques for digital images. In: 18th International Conference on Pattern Recognition (ICPR'06), vol. 4, pp. 358–361. IEEE (2006)
8. Kutter, M., Jordan, F.D., Bossen, F.: Digital watermarking of color images using amplitude modulation. J. Electron. Imaging **7**(2), 326–333 (1998)
9. Karmakar, A., Phadikar, A., Phadikar, B.S., Maity, G.K.: A blind video watermarking scheme resistant to rotation and collusion attacks. J. King Saud Univ. Comput. Informa. Sci. **28**(2), 199–210 (2016)
10. Kong, W., Yang, B., Wu, D., Niu, X.: SVD based blind video watermarking algorithm. In: First International Conference on Innovative Computing, Information and Control (ICICIC'06), vol. 1, pp. 265–268. IEEE (2006)
11. Liu, R., Tan, T.: An SVD-based watermarking scheme for protecting rightful ownership. IEEE Trans. Multimedia **4**(1), 121–128 (2002)
12. Singh, T.R., Singh, K.M., Roy, S.: Video watermarking scheme based on visual cryptography and scene change detection. AEU-Int. J. Electron. Commun. **67**(8), 645–651 (2013)
13. Faragallah, O.S.: Efficient video watermarking based on singular value decomposition in the discrete wavelet transform domain. AEU-Int. J. Electron. Commun. **67**(3), 189–196 (2013)
14. Kaur, M., Jindal, S., Behal, S.: A study of digital image watermarking. J. Res. Eng. Appl. Sci. **2**(2), 126–136 (2012)
15. Rasti, P., Samiei, S., Agoyi, M., Escalera, S., Anbarjafari, G.: Robust non-blind color video watermarking using QR decomposition and entropy analysis. J. Vis. Commun. Image Represent. **38**, 838–847 (2016)
16. Youssef, S.M., ElFarag, A.A., Ghatwary, N.M.: Adaptive video watermarking integrating a fuzzy wavelet-based human visual system perceptual model. Multimedia Tools Appl. **73**(3), 1545–1573 (2014)
17. Tabassum, T., Islam, S.M.: A digital video watermarking technique based on identical frame extraction in 3-Level DWT. In: 2012 15th International Conference on Computer and Information Technology (ICCIT), pp. 101–106. IEEE (2012)
18. Agilandeeswari, L., Ganesan, K.: A robust color video watermarking scheme based on hybrid embedding techniques. Multimedia Tools Appl. **75**(14), 8745–8780 (2016)
19. Xuemei, J., Quan, L., Qiaoyan, W.: A new video watermarking algorithm based on shot segmentation and block classification. Multimedia Tools Appl. **62**(3), 545–560 (2013)
20. Lu, C.S., Liao, H.Y., Sze, C.J.: Combined watermarking for image authentication and protection. In: 2000 IEEE International Conference on Multimedia and Expo, ICME2000. Proceedings. Latest Advances in the Fast Changing World of Multimedia (Cat. No. 00TH8532), vol. 3, pp. 1415–1418. IEEE (2000)
21. Li, C.T., Yang, F.M.: One-dimensional neighborhood forming strategy for fragile watermarking. J. Electron. Imaging **12**(2), 284–292 (2003)

22. Hernandez, J.R., Perez-Gonzalez, F., Rodriguez, J.M.: The impact of channel coding on the performance of spatial watermarking for copyright protection. In: Proceedings of the 1998 IEEE International Conference on Acoustics, Speech and Signal Processing, ICASSP'98 (Cat. No. 98CH36181), vol. 5, pp. 2973–2976. IEEE (1998)
23. Singh, A.K., Sharma, N., Dave, M., Mohan, A.: A novel technique for digital image watermarking in spatial domain. In: 2012 2nd IEEE International Conference on Parallel, Distributed and Grid Computing, pp. 497–501. IEEE (2012)
24. Singh, P., Chadha, R.: A survey of digital watermarking techniques, applications and attacks. Int. J. Eng. Innovative Technol. (IJEIT) **2**(9), 165–175 (2013)
25. Brannock, E., Weeks, M., Harrison, R.: Computer Science Department Georgia State University watermarking with wavelets: simplicity leads to robustness, Southeastcon (2008)

Printed in the United States
By Bookmasters